FRACTIONS

Connie Eichhorn

Power Math Series

CAMBRIDGE Adult Education
A Division of Simon & Schuster
Upper Saddle River, New Jersey

Dr. Connie Eichhorn is the Supervisor of Transitional Programs for the Omaha Public Schools. She is the former president of the American Association of Adult and Continuing Education. Dr. Eichhorn is very active in adult education and has consulted in the development of a variety of adult education materials.

EXECUTIVE EDITOR: Mark Moscowitz

EDITOR: Kirsten Richert

PRODUCTION DIRECTOR: Penny Gibson

PRODUCTION EDITOR: Linda Greenberg

PRINT BUYER: Patricia Alvarez

ART DIRECTOR: Marianne Frasco

BOOK DESIGN: Patrice Sheraton

ELECTRONIC PAGE PRODUCTION: Curriculum Concepts

COVER DESIGN: Amy Rosen

COVER PHOTO: © David Bishop/Phototake

Printed in the United States of America

1 2 3 4 5 6 7 8 9 10   99 98 97 96 95

ISBN 0-13-439936-6

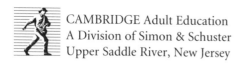

CAMBRIDGE Adult Education
A Division of Simon & Schuster
Upper Saddle River, New Jersey

Contents

To the Learner

The ten books in the Power Math series are designed to help you understand and practice arithmetic skills. Lessons are easy to use and the problems are designed to address every-day adult life.

Lessons have the following features:

- Every lesson begins with a sample problem from real-life experience. You are asked to use your knowledge of math to find a solution.

- The *Think* section takes you through the thought process you might use to organize the information in the problem and choose a problem-solving approach.

- The *Do* section shows you step-by-step how to solve the problem.

- In *Try These*, you will warm up by solving a few problems similar to the opening sample problem. Some steps are worked for you to get you off to a good start.

- The *Practice* section gives you ample opportunities to practice the skill presented in the lesson.

- The *Solving Problems* section applies the math skill in a practical application from your experience as a consumer and worker.

Within each book, review lessons give you opportunities to decide whether you have mastered the skills presented in the book. The *Answer Key* section at the end of the book has answers and worked-out solutions for the problems in the book. Use the answers to check your work. Use the worked-out solutions to make sure your approach to a problem was the correct one.

By working carefully through the exercises in this book, you will find increased confidence in your math skills. Good luck.

What Are Fractions?

A fraction is a part of a whole number or group of numbers. You use fractions every day of your life. When you split a piece of pie "in half" or share some money by dividing it "into thirds" you are using fractions.

The square on the right is divided into four equal parts or fourths. One of the parts is shaded.

How can you write a fraction to represent the shaded part?

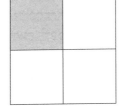

Think

Fractions are written using two numbers—a *numerator* and a *denominator*. The numerator shows how many parts of the whole are in the fraction. The denominator shows into how many parts the whole is divided.

The numerator is the top number. Write the numerator first, then draw a line under it. Below the line, write the denominator.

Sometimes you will see fractions written in words. For example, you will often read or hear one-half, two-thirds, and three-fourths. When a fraction is written in words, the first number is the numerator. The following words are often used to express denominators:

seconds, halves	2
thirds	3
fourths	4
fifths	5
sixths	6
sevenths	7
eighths	8
ninths	9
tenths	10

Do

The square is divided into four equal parts so the denominator is 4. To write a fraction that represents the shaded part:

Step 1. Write the number 1.

Step 2. Draw a line under the 1.

Step 3. Write the number 4 under the line.

The fraction is $\frac{1}{4}$. In words, you write "one fourth."

Try These

Write fractions as indicated.

1. What part of the circle is shaded? _____
    The numerator is 2.
    The denominator is 3.
    The fraction is $\frac{2}{?}$.

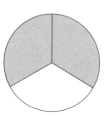

2. What part of the cake is decorated? _____
    The numerator is 5
    The denominator is ?.
    The fraction is $\frac{?}{?}$.

PRACTICE

Write a fraction to describe each of the following.

3. One part of a rope cut into two equal pieces. _____

4. Three slices of a pizza divided into eight equal slices. _____

5. Two teaspoons out of three teaspoons of flour. _____

6. Four cups out of 5 cups of water. _____

7. Three parts of a rectangle divided into four equal parts. _____

Write the fractions using numbers.

8. One-third _____

9. Two-fifths _____

10. One-half _____

11. Five-eighths _____

12. Nine-tenths _____

Solving Problems

Solve.

13. Write the fraction that shows the portion of the circle that is shaded.

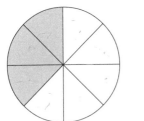

_____

14. Write the fraction that shows the portion of the circle that is *not* shaded.

_____

Check your answers on page 75.

2

Equivalent Fractions

Pat wants to eat $\frac{1}{3}$ of a pizza. The pizza is cut into six pieces. Pat eats two slices. Is $\frac{2}{6}$ of a pizza the same as $\frac{1}{3}$?

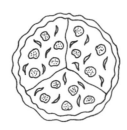

Think

Two fractions are called equivalent, or equal, if they represent the same part of a whole or group of numbers. You can know whether two fractions are equal by comparing their *cross products.*

You can find the cross products by multiplying the numerator of one fraction by the denominator of the other. Then multiply the other numerator by the other denominator.

Do

Is $\frac{1}{3} = \frac{2}{6}$?

Step 1. Multiply the numerator of $\frac{1}{3}$ (1) by the denominator of $\frac{2}{6}$ (6). $1 \times 6 = 6$

$$\frac{1}{3} \quad \frac{2}{6}$$

$3 \times 6 = 36$

Step 2. Multiply the denominator of $\frac{1}{3}$ (3) by the numerator of $\frac{2}{6}$ (2). $3 \times 2 = 6$

$$\frac{1}{3} \quad \frac{2}{6} \qquad 1 \times 6 = 2 \times 3$$

Since the cross products of $\frac{1}{3}$ and $\frac{2}{6}$ are equal, the two fractions are equivalent. Pat will eat $\frac{1}{3}$ of a pizza if she eats two of six slices.

Try These

Cross multiply to decide whether the fractions are equivalent.

1. $\frac{2}{4}$ and $\frac{4}{8}$

   $2 \times 8 = 16$
   $4 \times 4 = ?$

   Are the fractions equivalent? _____

2. $\frac{3}{5}$ and $\frac{5}{10}$

   $3 \times 10 = 30$
   $? \times ? = ?$

   Are the fractions equivalent? _____

--------- **PRACTICE** ---------

Cross multiply to decide whether the fractions are equivalent.

3. $\frac{3}{4}$ and $\frac{6}{8}$ _____ 4. $\frac{1}{5}$ and $\frac{3}{8}$ _____ 5. $\frac{1}{3}$ and $\frac{3}{9}$ _____

6. $\frac{4}{8}$ and $\frac{1}{2}$ _____ 7. $\frac{2}{5}$ and $\frac{4}{9}$ _____ 8. $\frac{2}{10}$ and $\frac{1}{5}$ _____

9. $\frac{2}{8}$ and $\frac{1}{4}$ _____ 10. $\frac{2}{3}$ and $\frac{4}{6}$ _____

Solving Problems

Solve.

11. Chris needs to get $\frac{1}{4}$ cup of water. He has a $\frac{1}{8}$ cup measuring spoon. Is $\frac{2}{8}$ cup equivalent to $\frac{1}{4}$ cup?

12. Juan has $\frac{3}{5}$ of a pound of apples. Maria has $\frac{5}{10}$ of a pound. Are the two weights equal?

_____   _____

Check your answers on page 75.

3

Raising to Higher Terms

Kim is cutting a watermelon. He plans to cut the watermelon into eighths. How many eighths will he be able to cut from one half of the watermelon?

Think

When you multiply the numerator and denominator of a fraction by the same whole number, you create an equivalent fraction. The terms or numbers in the new fraction are higher. This process is called *raising a fraction to higher terms*.

numerator $\times$ whole number = higher numerator

denominator $\times$ same whole number = higher denominator

The original fraction and the fraction with higher terms are equal. That means the two fractions represent equal parts of a whole or group of numbers.

Do

You are trying to find out how many eighths it will take to equal $\frac{1}{2}$ of the watermelon. You need to raise $\frac{1}{2}$ to a fraction with a denominator of 8.

$$\frac{1}{2} = \frac{?}{8}$$

Step 1. Multiply the numerator 1 by 4.

$1 \times 4 = 4$

Step 2. Multiply the denominator 2 by 4.

$2 \times 4 = 8$

$\frac{4}{8}$ is equivalent to $\frac{1}{2}$.

If Kim cuts the watermelon into eighths, four pieces will equal one half of the watermelon.

> **THINK**
>
> What number would you multiply by 2 to get 8?   4
>
> To raise the fraction to higher terms, multiply both the numerator and the denominator by 4.

Try These

Raise each of these fractions to higher terms by multiplying the numerators and denominators by the same whole number.

1. Find an equivalent fraction of $\frac{1}{3}$ by raising to a higher term. Multiply by 3.
   $1 \times 3 = 3$
   $3 \times 3 = 9$
   The equivalent fraction is $\frac{?}{?}$. _____ $\frac{9}{21}$

2. Find an equivalent fraction of $\frac{1}{4}$ by raising to a higher term. Multiply by 2.
   $1 \times 2 = 2$
   $? \times ? = ?$
   The equivalent fraction is $\frac{?}{?}$. _____

PRACTICE

Raise each of these fractions to higher terms by multiplying the numerators and denominators by the same whole number.

Fraction	Multiply By	Equivalent Fraction
3. $\frac{1}{2}$	2	_____
4. $\frac{1}{2}$	3	_____
5. $\frac{1}{2}$	4	_____
6. $\frac{1}{2}$	5	_____
7. $\frac{1}{3}$	2	_____
8. $\frac{2}{3}$	2	_____
9. $\frac{2}{3}$	3	_____
10. $\frac{3}{4}$	2	_____

Solving Problems

Raise each of these fractions to higher terms by multiplying the numerators and denominators by the same whole number.

11. Bill is looking for two equivalent fractions for $\frac{2}{3}$. He wants the fractions to have denominators of 6 and 9. What are the fractions?

    _____

12. Gloria has $\frac{1}{4}$ of a pound of chicken. What is the equivalent fraction in eighths?

    _____

Check your answers on page 75.

Carla has $\frac{4}{8}$ of a cup of water. She wants to find an equivalent fraction with a smaller denominator.

Think

Equivalent fractions can also be made by dividing the numerator and denominator by the same whole number. *Reducing* or *simplifying* a fraction means finding an equivalent fraction with a lower denominator.

You have reduced a fraction to *lowest terms* when the only whole number you can divide both the numerator and denominator by is 1.

If a fraction has even numbers (2, 4, 6, 8, and so on) in both the numerator and denominator, you can divide them by 2 to reduce the fraction. If both numerator and denominator are not even, check to see if you can divide them both by 3 or 5. Remember, you must divide the numerator and denominator by the *same* whole number.

Do

To reduce the fraction $\frac{4}{8}$ to the lowest terms, Carla notes that both the numerator and denominator are even numbers. She divides both by 2.

$$\frac{4 \div 2 = 2}{8 \div 2 = 4}$$

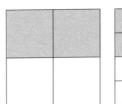 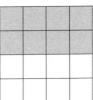

The fraction $\frac{2}{4}$ is reduced from $\frac{4}{8}$. Carla sees that both the numerator and denominator of $\frac{2}{4}$ are even numbers also. Carla divides them by 2 again.

$$\frac{2 \div 2 = 1}{4 \div 2 = 2}$$

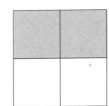 

$\frac{1}{2}$ is reduced from $\frac{2}{4}$. Since 1 is the only whole number that divides evenly into the numerator and denominator, $\frac{1}{2}$ is in lowest terms.

Try These

Reduce each of these fractions to their lowest terms.

1. $\frac{4}{6}$

$\frac{4 \div 2 = 2}{6 \div 2 = ?}$

In lowest terms, $\frac{4}{6}$ equals $\frac{?}{?}$. _____

2. $\frac{6}{9}$

$\frac{6 \div ? = 2}{9 \div 3 = ?}$

In lowest terms, $\frac{6}{9}$ equals $\frac{?}{?}$. _____

PRACTICE

Reduce each of these fractions to lowest terms.
(NOTE: Some of the fractions cannot be reduced.)

3. $\frac{2}{8}$ _____

4. $\frac{3}{6}$ _____

5. $\frac{7}{8}$ _____

6. $\frac{6}{8}$ _____

7. $\frac{4}{10}$ _____

8. $\frac{8}{10}$ _____

9. $\frac{3}{5}$ _____

10. $\frac{2}{6}$ _____

Solving Problems

Solve

11. Carlos travels $\frac{2}{10}$ of a mile. Can he reduce the fraction to lower terms? If so, write the fraction.

12. Lisa needs $\frac{5}{8}$ ounces of chocolate to make dessert. Can she reduce the fraction to lower terms? If so, write the fraction.

_____

_____

Check your answers on page 75.

LESSON

5

Adding Like Fractions

Hakim eats a $\frac{1}{4}$ pound cheeseburger. He is still hungry, so he orders another $\frac{1}{4}$ pound cheeseburger. What is the total weight of the meat on the cheeseburgers Hakim ate?

Think

You need to add $\frac{1}{4} + \frac{1}{4}$ to find the total amount Hakim ate.

Like fractions are fractions that have the same denominator. To add like fractions, add the numerators and use the like denominator. The answer should then be reduced to its lowest terms.

Do

Add: $\frac{1}{4} + \frac{1}{4}$. You know these are like fractions because the denominators are the same.

Step 1. Add the numerators.

$1 + 1 = 2$

Step 2. Use the like denominator.   4

The sum of $\frac{1}{4} + \frac{1}{4} = \frac{2}{4}$.

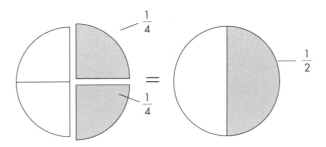

Step 3. You can reduce the fraction to its lowest terms by dividing the numerator and denominator by 2.

$$\frac{2 \div 2 = 1}{4 \div 2 = 2}$$

In lowest terms, Hakim ate a $\frac{1}{2}$ pound of meat.

11

Try These

Add these like fractions. Reduce the answer to lowest terms, if necessary.

1. $\frac{2}{5} + \frac{1}{5}$

   $2 + 1 = ?$

   The sum is $\frac{?}{5}$. _____

2. $\frac{1}{10} + \frac{3}{10}$

   $1 + 3 = ?$

   The sum is $\frac{4}{10}$. _____
   Reduced to lowest
   terms, the sum is $\frac{?}{?}$. _____

PRACTICE

Add these like fractions. Reduce the answer to lowest terms, if necessary.

3. $\frac{1}{3} + \frac{1}{3}$ _____

4. $\frac{1}{4} + \frac{2}{4}$ _____

5. $\frac{3}{8} + \frac{1}{8}$ _____

6. $\frac{1}{5} + \frac{3}{5}$ _____

7. $\frac{2}{6} + \frac{3}{6}$ _____

8. $\frac{4}{8} + \frac{2}{8}$ _____

9. $\frac{2}{9} + \frac{4}{9}$ _____

10. $\frac{2}{4} + \frac{1}{4}$ _____

Solving Problems

Solve. Reduce to lowest terms, if necessary.

11. Dave is building a wood cabinet. He must add $\frac{3}{8}$ of an inch and $\frac{1}{8}$ of an inch. What is the sum?

12. Toni cuts a length of rope into 6 equal pieces. She picks up $\frac{2}{6}$ of the rope and then another $\frac{3}{6}$. How much of the original rope is she holding?

_____   _____

Check your answers on page 76.

Subtracting Like Fractions

Jeff cuts an orange into eighths. He puts $\frac{5}{8}$ of the orange on a plate and offers some to his sister. She takes 3 pieces. How much of the orange is left on the plate?

Think

Jeff must subtract $\frac{5}{8} - \frac{3}{8}$ to find out how much he has left.

Subtracting like fractions is as easy as adding like fractions. Just subtract the numerators and use the like denominator.

After subtracting the fractions, reduce the answer to lowest terms, if necessary.

Do

$$\frac{5}{8} - \frac{3}{8}$$

Step 1. Subtract the numerators. $\qquad 5 - 3 = 2$

Step 2. Use the like denominator. $\qquad 8$
The result is $\frac{2}{8}$.

Step 3. Reduce the result. $\qquad \dfrac{2 \div 2 = 1}{8 \div 2 = 4}$

Jeff has $\frac{1}{4}$ of an orange left.

Try These

Subtract these like fractions. Reduce the answer, if necessary.

1. $\dfrac{4}{6} - \dfrac{1}{6}$

$4 - 1 = 3$
The difference is $\frac{3}{?}$.

Reduced to lowest terms, the answer is $\frac{?}{?}$ _____

2. $\dfrac{6}{9} - \dfrac{3}{9}$

$6 - ? = ?$

The difference is $\dfrac{?}{9}$.

Reduced to lowest terms, the answer is $\dfrac{?}{?}$. _____

PRACTICE

Subtract these like fractions. Reduce the answer, if necessary.

3. $\dfrac{4}{8} - \dfrac{2}{8}$ _____

4. $\dfrac{3}{4} - \dfrac{2}{4}$ _____

5. $\dfrac{4}{5} - \dfrac{2}{5}$ _____

6. $\dfrac{7}{8} - \dfrac{3}{8}$ _____

7. $\dfrac{5}{6} - \dfrac{2}{6}$ _____

8. $\dfrac{8}{9} - \dfrac{5}{9}$ _____

9. $\dfrac{7}{10} - \dfrac{3}{10}$ _____

10. $\dfrac{2}{3} - \dfrac{1}{3}$ _____

Solving Problems

Solve. Reduce the answer, if necessary.

11. Ann cuts $\dfrac{7}{8}$ of a yard of wood. From that piece, she saws off $\dfrac{5}{8}$ of a yard. How much wood is left?

12. Mark has $\dfrac{9}{10}$ of an ounce of mushrooms. He uses $\dfrac{3}{10}$ of an ounce of mushrooms cooking dinner. What part of an ounce does Mark have left?

_____

_____

Check your answers on page 76.

Finding a Common Denominator

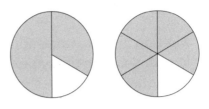

Ted has to add the fractions $\frac{1}{2} + \frac{1}{3}$. He knows how to add like fractions, but these are not like fractions. How can Ted rewrite the fractions so that the denominators are the same?

Think

Before fractions can be added, they must be like fractions. If the fractions to be added are not like fractions, one or the other, or both, must be raised to higher terms.

Before Ted can add $\frac{1}{2}$ and $\frac{1}{3}$, he needs to think of a denominator that both $\frac{1}{2}$ and $\frac{1}{3}$ could be raised to. He needs to *find a common denominator.*

> **REMEMBER**
>
> To raise a fraction to higher terms, you have to multiply the numerator and the denominator by the same whole number.

To find a common denominator:

See if you can raise the fraction with the lower denominator to the same denominator as the other fraction. For example, if your two fractions are $\frac{1}{8}$ and $\frac{1}{4}$, you can raise $\frac{1}{4}$ to eighths by multiplying the numerator and denominator by 2.

$$\frac{1}{4} \times \frac{2}{2} = \frac{2}{8}$$

Try to think of a number that both denominators will divide into. For example, if your fractions are $\frac{1}{4}$ and $\frac{1}{6}$, you can use 12 as a denominator to make like fractions.

$$\frac{1}{4} \times \frac{3}{3} = \frac{3}{12} \qquad\qquad \frac{1}{6} \times \frac{2}{2} = \frac{2}{12}$$

The fractions $\frac{3}{12}$ and $\frac{2}{12}$ are equivalent fractions of $\frac{1}{4}$ and $\frac{1}{6}$.

If you have trouble finding a common denominator, you can always multiply the two denominators. This always results in a common denominator. For example, if your two fractions are $\frac{2}{3}$ and $\frac{2}{5}$, you can use 15 (5 × 3) as the common denominator.

$$\frac{2 \times 5 = 10}{5 \times 5 = 15}$$

$$\frac{2 \times 3 = 6}{5 \times 3 = 15}$$

The fractions $\frac{10}{15}$ and $\frac{6}{15}$ have common denominators.

Multiplying the two denominators always works, but the result may not be the lowest common denominator. After multiplying, try dividing the common denominator by 2, 3, or 5 to see if you can find the lowest common denominator.

Do

Ted must find a common denominator for $\frac{1}{2}$ and $\frac{1}{3}$. He decides to use 6. He found this by multiplying the denominators 2 and 3. He cannot think of any lower common denominator.

$$\frac{1 \times 3 = 3}{2 \times 3 = 6}$$

$$\frac{1 \times 2 = 2}{3 \times 2 = 6}$$

Now Ted has two fractions, $\frac{3}{6}$ and $\frac{2}{6}$, with common denominators.

Try These

Find the least common denominator for these pairs of fractions.

1. $\frac{3}{4}$ and $\frac{3}{8}$ _____

   You should multiply the lower denominator, 4, by _____ to equal the higher denominator, 8.

2. $\frac{5}{6}$ and $\frac{1}{4}$ _____

   You can multiply 6 × 4 to get the common denominator 24. You can divide by ___2___ to find the lowest common denominator, which is _____.

PRACTICE

Find the least common denominator for these pairs of fractions.

3. $\frac{2}{3}$ and $\frac{1}{4}$ _____

4. $\frac{3}{8}$ and $\frac{1}{2}$ _____

5. $\frac{1}{2}$ and $\frac{1}{3}$ _____

6. $\frac{1}{2}$ and $\frac{3}{4}$ _____

7. $\frac{2}{5}$ and $\frac{1}{2}$ _____

8. $\frac{1}{5}$ and $\frac{2}{3}$ _____

9. $\frac{1}{4}$ and $\frac{3}{5}$ _____

10. $\frac{1}{6}$ and $\frac{3}{4}$ _____

Solving Problems

Solve.

11. Sal has two pieces of lumber. One is $\frac{3}{4}$ of a foot; the other is $\frac{5}{8}$ of a foot. Before he can add the two fractions, Sal must make them into like fractions. What is the least common denominator for the two fractions?

12. Alicia wants to find the sum of $\frac{3}{4}$ and $\frac{2}{5}$. What is the least common denominator for the two fractions?

_____   _____

Check your answers on page 76.

Adding Unlike Fractions

Barbara has $\frac{1}{3}$ of a cup of flour in one bag and $\frac{1}{2}$ of a cup in another bag. She wants to add the two amounts to find out how much flour she has.

Think

Before you can add fractions, they must have a common denominator. Find the common denominator and raise one or both fractions to higher terms. Then you can add the like fractions.

Find the least common denominator.

Raise one or both fractions to higher terms so they have the common denominator.

Add the fractions together.

If you have used a common denominator that is not the *least* common denominator, you will have to reduce the answer to the lowest terms.

Do

Step 1. $\dfrac{1}{2} + \dfrac{1}{3}$

Find the least common denominator.

6

Step 2. Raise the fractions to higher terms so they both have the common denominator.

$$\dfrac{1 \times 3}{2 \times 3} = \dfrac{3}{6} \qquad\qquad \dfrac{1 \times 2}{3 \times 2} = \dfrac{2}{6}$$

Step 3. Add the fractions.

$$\dfrac{3}{6} + \dfrac{2}{6} = \dfrac{5}{6}$$

Barbara has a total of $\frac{5}{6}$ of a cup of flour.

Try These

Add these fractions after finding the least common denominator. Reduce the answer to lowest terms, if necessary.

1. $\dfrac{1}{2} + \dfrac{1}{4}$

    Find the least common denominator.

    4

    Raise the first fraction to higher terms using the common denominator.

    $\dfrac{1 \times 2}{2 \times 2} = \dfrac{2}{4}$

    Add the like fractions.

    $\dfrac{2}{4} + \dfrac{1}{4} = \dfrac{?}{?}$ _____

2. $\dfrac{2}{3} + \dfrac{1}{4}$

    Find the common denominator.

    12

    Raise the fractions to higher terms using the common denominator.

    $\dfrac{2 \times 4}{3 \times 4} = \dfrac{8}{?}$ $\dfrac{1 \times 3}{4 \times 3} = \dfrac{3}{?}$

    Add the like fractions.

    $\dfrac{8}{?} + \dfrac{3}{?} = \dfrac{?}{?}.$ _____

PRACTICE

Add these fractions after finding the least common denominator. Reduce the answer to lowest terms, if necessary.

3. $\dfrac{1}{3} + \dfrac{1}{4}$ _____

4. $\dfrac{1}{2} + \dfrac{2}{5}$ _____

5. $\dfrac{1}{4} + \dfrac{1}{5}$ _____

6. $\dfrac{1}{6} + \dfrac{1}{3}$ _____

7. $\dfrac{1}{4} + \dfrac{3}{8}$ _____

8. $\dfrac{1}{8} + \dfrac{2}{3}$ _____

9. $\dfrac{3}{8} + \dfrac{1}{2}$ _____

10. $\dfrac{2}{5} + \dfrac{1}{6}$ _____

Solving Problems

Solve.

11. Benito has $\frac{3}{4}$ pounds of apples. His mother gives him another $\frac{1}{8}$ of a pound. How many pounds of apples does Benito have now?

_____

12. Hariko walks $\frac{1}{5}$ of a mile. Then she walks $\frac{2}{3}$ of a mile more. How far did Hariko walk?

_____

Check your answers on page 76.

Subtracting Unlike Fractions

Betty has $\frac{1}{3}$ of her birthday cake left over. She gives her son $\frac{1}{6}$ of the cake for dessert. Betty wants to know how much of the cake remains.

Think

To subtract, you need like fractions. You need to find a common denominator, raise one or both fractions to higher terms, and then subtract the like fractions. If you have not used the *least* common denominator, you will have to reduce the answer to lowest terms.

Do

Step 1. $\frac{1}{3} - \frac{1}{6}$

Find the least common denominator.   6

Step 2. Raise the first fraction to higher terms so both fractions have the common denominator.

$$\frac{1}{3} \times \frac{2}{2} = \frac{2}{6}$$

Step 3. Subtract the fractions.

$$\frac{2}{6} - \frac{1}{6} = \frac{1}{6}$$

Betty has $\frac{1}{6}$ of the birthday cake left.

Try These

Subtract these fractions after finding the least common denominator.

1. $\frac{3}{4} - \frac{1}{8}$

Find the least common denominator.   8

Raise the terms of the fractions to the common denominator.

$$\frac{3}{4} \times \frac{2}{2} = \frac{6}{8}$$

Subtract the fractions.

$$\frac{6}{8} - \frac{1}{8} = \frac{?}{8}. \quad \underline{\qquad}$$

2. $\dfrac{1}{3} - \dfrac{1}{4}$

Find the least common denominator. $\qquad$ 12

Raise the terms of the fractions to the common denominator.

$$\dfrac{1 \times 4}{3 \times 4} = \dfrac{?}{12} \qquad \dfrac{1 \times 3}{4 \times 3} = \dfrac{3}{?}$$

Subtract the fractions.

$$\dfrac{?}{12} - \dfrac{3}{?} = \dfrac{?}{?}. \ \underline{\qquad}$$

PRACTICE

Subtract these fractions after finding a least common denominator. Reduce the answer to lowest terms, if necessary.

3. $\dfrac{1}{2} - \dfrac{1}{4}$ $\underline{\qquad}$

4. $\dfrac{2}{3} - \dfrac{1}{6}$ $\underline{\qquad}$

5. $\dfrac{5}{8} - \dfrac{1}{4}$ $\underline{\qquad}$

6. $\dfrac{7}{8} - \dfrac{3}{4}$ $\underline{\qquad}$

7. $\dfrac{2}{3} - \dfrac{1}{4}$ $\underline{\qquad}$

8. $\dfrac{3}{5} - \dfrac{1}{3}$ $\underline{\qquad}$

9. $\dfrac{3}{4} - \dfrac{2}{5}$ $\underline{\qquad}$

10. $\dfrac{2}{3} - \dfrac{5}{8}$ $\underline{\qquad}$

Solving Problems

Solve.

11. James has $\frac{3}{4}$ of a cup of milk left in a carton. He pours $\frac{1}{8}$ of a cup into a glass. How much milk does James have left in the carton?

12. Suji buys $\frac{5}{8}$ of a pound of cream cheese. She uses $\frac{1}{3}$ of a pound to make cheese cake. How much cream cheese does Suji have left?

$\underline{\qquad}$ $\qquad$ $\underline{\qquad}$

Check your answers on pages 76-77.

10

Changing Mixed Numbers to Improper Fractions

Kevin ran around the track $2\frac{1}{4}$ times. He wants to know how many quarter laps he ran.

Think

When the numerator is less than the denominator, the fraction is called a *proper fraction*. Proper fractions show a part of a whole that is less than 1.

Improper fractions stand for values of 1 or more. The numerator of an improper fraction is the same or greater than the denominator. $\frac{3}{2}$ and $\frac{4}{4}$ are improper fractions.

Mixed numbers also stand for values of 1 or more. Mixed numbers have two parts: a whole number and a proper fraction. $1\frac{1}{2}$ and $3\frac{3}{4}$ are mixed numbers.

You can think of a mixed number as adding a whole number to a fraction. When you say a mixed number aloud, say "and" between the whole number and the proper fraction. $1\frac{1}{2}$ is read "one and one half."

Kevin ran $2\frac{1}{4}$ laps. That amount is a mixed number. Kevin wants to figure out how many quarter laps he ran: $\frac{?}{4}$. He needs to change a mixed number ($2\frac{1}{4}$) to an improper fraction.

To turn a mixed number into an improper fraction:

Multiply the whole number part of the mixed number by the denominator of the fraction.

Add the result to the numerator of the fraction part. This becomes the numerator of the improper fraction.

The denominator remains the same.

Do

Change $2\frac{1}{4}$ to an improper fraction.

Step 1. Multiply the whole number part (2) by the denominator of the fraction part (4).

$2 \times 4 = 8$

Step 2. Add the result to the numerator of the fraction part (1).

$8 + 1 = 9$

Step 3. Use the result (9) as the numerator. The denominator is still 4.

Kevin ran $\frac{9}{4}$ laps around the track.

Try These

Write each mixed number as an improper fraction.

1. $3\frac{3}{8} = $ _____

   Multiply the whole number part by the denominator of the fraction part.

   $3 \times 8 = 24$

   Add the result to the numerator of the fraction part.

   $24 + 3 = 27$

   Write the result over the denominator.

   $3\frac{3}{8} = \frac{27}{?}$

2. $4\frac{2}{3} = $

   Multiply: $4 \times 3 = 12$

   Add: $12 + $ _____ $= $ _____

   Write the improper fraction: $4\frac{2}{3} = $ _____

Write these mixed numbers as improper fractions.

3. $2\frac{1}{4}$ ____

4. $1\frac{5}{8}$ ____

5. $3\frac{2}{3}$ ____

6. $2\frac{3}{5}$ ____

7. $8\frac{1}{6}$ ____

8. $6\frac{2}{5}$ ____

9. $1\frac{3}{4}$ ____

10. $5\frac{7}{8}$ ____

Solving Problems

Solve.

11. Ricardo bought $3\frac{2}{3}$ pounds of potatoes. How can he write the amount as an improper fraction?

12. John spent $2\frac{2}{3}$ hours waiting to take his driver's license test. How can he write this amount as an improper fraction?

_____ _____

Check your answers on page 77.

L E S S O N

Changing
Improper
Fractions to
Mixed
Numbers

Gabe worked $\frac{3}{4}$ of an hour of overtime on Monday and $\frac{3}{4}$ of an hour on Tuesday. When he added the fractions together, he found he had worked $\frac{6}{4}$ hours of overtime. How can he write the amount of time as a mixed number?

Think

Improper fractions can be changed to mixed numbers by dividing the numerator by the denominator. When you divide, the answer becomes the whole number part of the mixed number. The remainder becomes the numerator of the fraction part. The denominator remains the same.

Do

To turn the improper fraction $\frac{6}{4}$ into a mixed number:

Step 1. Divide the numerator (6) by the denominator (4).

$$\begin{array}{r} 1 \\ 4\overline{)6} \\ \underline{4} \\ 2 \end{array}$$

Step 2. The whole number result (1) is the whole number part of the mixed number.

Step 3. The remainder (2) becomes the numerator of the fraction part. The denominator is 4, the number you divided by.

The improper fraction $\frac{6}{4}$ equals the mixed number $1\frac{2}{4}$.

Reduce $1\frac{2}{4}$ to lowest terms.

$$1\frac{2}{4} = 1\frac{1}{2}$$

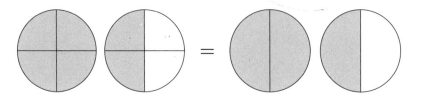

Try These

Turn this improper fraction into a mixed number.

1. $\dfrac{14}{3}$ Divide the numerator by the denominator. $14 \div 3 = $ _____

The whole number result is the whole number part of the mixed number. $14 \div 3 = 4 \ R2$

The remainder is the numerator of the fraction part. The denominator remains the same.

$\dfrac{14}{3} = 4\dfrac{2}{?}$ _____

2. $\dfrac{12}{3} = $ Divide: $12 \div 3 = ?$

There is no remainder, so the improper fraction equals the whole number _____.

PRACTICE

Turn these improper fractions into mixed numbers. Reduce to lowest terms if necessary.

3. $\dfrac{5}{2}$ _____

4. $\dfrac{7}{3}$ _____

5. $\dfrac{10}{4}$ _____

6. $\dfrac{9}{3}$ _____

7. $\dfrac{7}{2}$ _____

8. $\dfrac{8}{3}$ _____

9. $\dfrac{12}{5}$ _____

10. $\dfrac{23}{4}$ _____

Solving Problems

Solve.

11. After a party, Sokry noticed she had 17 slices of pizza left over. Each of the pizzas she ordered had been cut into 8 slices. She knows she can write the amount of leftover pizza as the improper fraction $\frac{17}{8}$. How could she write that amount as a mixed number?

12. Lisa has 21 quarter-ounce pieces of chocolate. The amount can be written $\frac{21}{4}$. How can Lisa write this amount as a mixed number?

_____

_____

Check your answers on page 77.

Adding Mixed Numbers With Like Fractions

Terry drives $2\frac{3}{4}$ miles to pick up his son from school. Then he takes his son home, driving another $1\frac{3}{4}$ miles. How many total miles did Terry drive?

Think

Terry sees that the distances he drove are mixed numbers. He thinks of them as "two and three quarters" and "one and three quarters" miles.

To add mixed numbers, separate the problem into two additions.

Add the whole number parts of the mixed numbers.

Add the fraction parts of the mixed number.

If you get an improper fraction, turn it into a mixed number and add the whole number part to your answer.

Reduce the fraction to lowest terms, if necessary.

Do

$$2\frac{3}{4} + 1\frac{3}{4}$$

Step 1. Add the whole number parts of the mixed numbers.

$$2 + 1 = 3$$

Step 2. Add the fraction parts of the mixed numbers.

$$\frac{3}{4} + \frac{3}{4} = \frac{6}{4}$$

Step 3. Turn the improper fraction into a mixed number.

$$\frac{6}{4} = 1\frac{2}{4}$$

Step 4. Add the whole number of this mixed fraction to the whole number you got in step 1.

$$3 + 1\frac{2}{4} = 4\frac{2}{4}$$

Step 5. Reduce the fraction to lowest terms.

$$4\frac{2}{4} = 4\frac{1}{2}$$

Terry drove a total of $4\frac{1}{2}$ miles.

Try These

Add these mixed numbers. Reduce the fraction part to lowest terms, if necessary.

1. $3\frac{1}{5} + 2\frac{2}{5} =$

   $3 + 2 = 5$

   $\frac{1}{5} + \frac{2}{5} = \frac{?}{5}$

   The answer is ? $\frac{?}{5}$ _____

2. $1\frac{7}{8} + 4\frac{5}{8}$

   $1 + ? = ?$

   $\frac{7}{8} + \frac{5}{8} = \frac{12}{8}$

   $\frac{12}{8}$ as a mixed number is ? $\frac{?}{8}$

   $\frac{?}{8}$ reduced to lowest terms is $\frac{?}{?}$

   The answer is ? $\frac{?}{?}$ _____

PRACTICE

Add these mixed numbers. Reduce the fraction part to lowest terms, if necessary.

3. $3\frac{1}{3} + 2\frac{1}{3}$ _____

4. $1\frac{1}{4} + 2\frac{1}{4}$ _____

5. $2\frac{1}{6} + 3\frac{2}{6}$ _____

6. $3\frac{1}{4} + 2\frac{2}{4}$ _____

7. $1\frac{1}{5} + 1\frac{3}{5}$ _____

8. $4\frac{1}{8} + 3\frac{5}{8}$ _____

9. $2\frac{2}{5} + 2\frac{4}{5}$ _____

10. $5\frac{5}{6} + 3\frac{5}{6}$ _____

Solving Problems

Solve.

11. Samir has two cans of diced tomatoes. One can holds $1\frac{1}{8}$ cups and the other can holds $2\frac{3}{8}$ cups. How many cups of tomatoes does Samir have?

12. Karen has two wooden dowels. One is $5\frac{7}{8}$ feet long. The other is $3\frac{3}{8}$ feet. What is the total length of the dowels?

_____

_____

Check your answers on pages 77–78.

13

Adding Mixed Numbers With Unlike Fractions

Mohammad is studying for a big test on Wednesday. On Monday he studies for $1\frac{3}{4}$ hours. On Tuesday he studies for another $2\frac{1}{2}$ hours. Mohammad wonders how many hours he studied on the two days.

Think

Mohammad sees that he must add the two mixed numbers. But he notices that the fraction parts are not like fractions. Before he can add the fraction parts, Mohammad must find a common denominator and raise one or both of the fractions to higher terms. Then he can add the mixed numbers with like fractions.

Do

$$1\frac{3}{4} + 2\frac{1}{2}$$

Step 1. Find the least common denominator for $\frac{3}{4}$ and $\frac{1}{2}$.
Since $2 \times 2 = 4$, the higher denominator is the least common denominator.

Step 2. Raise $\frac{1}{2}$ to higher terms.

$$\frac{1}{2} \times \frac{2}{2} = \frac{2}{4}$$

$$\frac{1}{2} = \frac{2}{4}$$

Step 3. Add the whole number parts.
$$1 + 2 = 3$$

Step 4. Add the fraction parts.
$$\frac{3}{4} + \frac{2}{4} = \frac{5}{4}$$

Step 5. Since $\frac{5}{4}$ is an improper fraction, change it to a mixed number.
$$\frac{5}{4} = 1\frac{1}{4}$$

Step 6. Add the whole number part.

$$3 + 1\frac{1}{4} = 4\frac{1}{4}$$

Mohammad studied for $4\frac{1}{4}$ hours on the two days.

Try These

Add these mixed numbers with unlike fractions. Reduce to lowest terms, `if necessary.

1. $1\frac{1}{3} + 2\frac{1}{2}$

    Change the fraction parts to like fractions.

    $$\frac{1}{3} \times \frac{2}{2} = \frac{2}{6}$$

    $$\frac{1}{2} \times \frac{3}{3} = \frac{3}{6}$$

    Add the whole number part.

    $$1 + 2 = 3$$

    Add the like fractions.

    $$\frac{2}{6} + \frac{3}{6} = \frac{?}{6}$$

    The answer is $3\frac{?}{6}$ _____

2. $2\frac{1}{2} + 1\frac{3}{4}$

    Change the fraction parts to like fractions.

    $$\frac{1}{2} \times \frac{2}{2} = \frac{2}{4}$$

    Add the whole number parts.

    $$2 + ? = ?$$

    Add the like fractions.

    $$\frac{2}{4} + \frac{3}{4} = \frac{5}{4}$$

    Turn the improper fraction into a mixed number.

    $$\frac{5}{4} = 1\frac{?}{4}$$

    Add the whole number.

    $$? + 1\frac{?}{4} = ?$$

    The answer is $?\frac{?}{4}$. _____

Add these mixed numbers with unlike fractions. Reduce to lowest terms, if necessary.

3. $2\frac{1}{3} + 1\frac{1}{2}$ _____

4. $1\frac{1}{8} + 1\frac{1}{4}$ _____

5. $3\frac{1}{2} + 1\frac{3}{4}$ _____

6. $1\frac{1}{3} + 2\frac{5}{6}$ _____

7. $2\frac{3}{8} + 1\frac{1}{2}$ _____

8. $1\frac{1}{3} + 2\frac{1}{4}$ _____

9. $2\frac{3}{4} + 3\frac{7}{8}$ _____

10. $1\frac{1}{4} + 2\frac{5}{6}$ _____

Solving Problems

Solve. Add these mixed numbers with unlike fractions. Reduce to lowest terms, if necessary.

11. Sarah makes $3\frac{1}{2}$ pounds of cookies for the first day of the school bake sale. She makes $2\frac{3}{4}$ pounds for the second day. How many pounds of cookies did Sarah bake?

12. Pete walked $1\frac{1}{2}$ miles to his friend's house. He walked $2\frac{1}{5}$ miles back, using a longer route. How many miles did Pete walk?

_____

_____

Check your answers on page 78.

Subtracting Mixed Numbers With Like Fractions

Holly bought $2\frac{3}{4}$ yards of fabric to make an outfit for her daughter. She uses $1\frac{1}{4}$ yards to make her daughter a skirt. How many yards of the fabric does Holly have left?

Think

You need to subtract $1\frac{1}{4}$ from $2\frac{3}{4}$. You will be working with mixed numbers with like fractions.
First, subtract the whole number parts.
Then, subtract the like fractions.
Reduce the fraction to lowest terms, if necessary.

Do

$2\frac{3}{4} - 1\frac{1}{4}$

Step 1. Subtract the whole number parts.

$2 - 1 = 1$

Step 2. Subtract the fraction parts.

$\frac{3}{4} - \frac{1}{4} = \frac{2}{4}$

Step 3. Reduce the fraction.

$\frac{2 \div 2}{4 \div 2} = \frac{1}{2}$

Holly has $1\frac{1}{2}$ yards of fabric left.

Try These

Subtract these mixed numbers with like fractions. Reduce your answers, if necessary.

1. $3\frac{5}{6} - 2\frac{1}{6}$

Subtract the whole number parts.

$3 - 2 = 1$

Subtract the fraction parts.

$\frac{5}{6} - \frac{1}{6} = \frac{4}{6}$

Reduce the fraction.

$\frac{4}{6} = \frac{?}{?}$

The answer is $1\frac{?}{?}$. _____

2. $5\frac{4}{5} - 3\frac{3}{5}$

Subtract the whole number parts.

$5 - ? = ?$

Subtract the fraction parts.

$\frac{4}{5} - \frac{3}{5} = \frac{?}{5}$

The answer is $?\frac{?}{5}$. _____

PRACTICE

Subtract these mixed numbers with like fractions. Reduce your answers, if necessary.

3. $2\frac{3}{4} - 1\frac{1}{4}$ _____

4. $3\frac{5}{6} - 1\frac{1}{6}$ _____

5. $4\frac{7}{8} - 2\frac{3}{8}$ _____

6. $4\frac{3}{5} - 3\frac{2}{5}$ _____

7. $5\frac{7}{9} - 2\frac{4}{9}$ _____

8. $6\frac{2}{3} - 5\frac{1}{3}$ _____

9. $5\frac{5}{8} - 1\frac{3}{8}$ _____

10. $8\frac{3}{4} - 5\frac{1}{4}$ _____

Solving Problems

Solve.

11. Melissa has $8\frac{4}{5}$ gallons of gasoline in her car. She uses $2\frac{2}{5}$ gallons driving to work. How many gallons of gasoline are left in Melissa's car?

12. Phil cooks a $5\frac{3}{4}$ pound roast. The family eats $3\frac{1}{4}$ pounds of it. How much of the roast is left over?

_____

_____

Check your answers on page 79.

Subtracting Mixed Numbers With Unlike Fractions

Pedro buys $3\frac{7}{8}$ pounds of soil to plant flowers. He uses $2\frac{3}{4}$ pounds of the soil to fill two large pots. Pedro wants to find out how much of the soil he has left.

Think

Pedro knows how to subtract mixed numbers with like fractions. But these numbers have unlike fractions. Before Pedro can subtract the mixed numbers, he has to turn the fractions to like fractions.

He will need to find a common denominator and raise one or both fractions to higher terms. Then he can subtract mixed numbers with like fractions.

Do

$$3\frac{7}{8} - 2\frac{3}{4}$$

Step 1. The least common denominator for $\frac{7}{8}$ and $\frac{3}{4}$ is 8. You can see this because 4 can be multiplied by 2 to get 8.

Step 2. Change $\frac{3}{4}$ to eighths by raising terms. Remember, to raise terms, multiply both the numerator and the denominator by the same whole number.

$$\frac{3}{4} \times \frac{2}{2} = \frac{6}{8}$$

Step 3. Subtract the whole number parts.

$$3 - 2 = 1$$

Step 4. Subtract the fraction parts.

$$\frac{7}{8} - \frac{6}{8} = \frac{1}{8}$$

Pedro has $1\frac{1}{8}$ pounds of soil remaining.

Try These

Subtract these mixed numbers with unlike fractions. Reduce your answers, if necessary.

1. $3\frac{5}{6} - 1\frac{1}{3}$

The least common denominator is 6. Change the fractions to like fractions.

$$\frac{1}{3} \times \frac{2}{2} = \frac{2}{6}$$

Subtract the whole number parts.

$$3 - 1 = 2$$

Subtract the like fractions.

$$\frac{5}{6} - \frac{2}{6} = \frac{3}{6}$$

Reduce the fraction, if necessary.

$$\frac{3}{6} = \frac{?}{?}$$

The answer is $2\frac{?}{?}$. _____

2. $5\frac{2}{3} - 2\frac{1}{4}$

The least common denominator is 12. Change the fraction parts to like fractions.

$$\frac{2 \times 4}{3 \times 4} = \frac{8}{12} \qquad \frac{1 \times 3}{4 \times 3} = \frac{3}{12}$$

Subtract the whole number parts.

$$5 - ? = ?$$

Subtract the like fractions.

$$\frac{8}{12} - \frac{3}{12} = \frac{?}{12}$$

The answer is $?\frac{?}{12}$. _____

PRACTICE

Subtract these mixed numbers with unlike fractions. Reduce your answers, if necessary.

3. $1\frac{1}{2} - 1\frac{1}{4}$ _____

4. $2\frac{5}{6} - 1\frac{2}{3}$ _____

5. $3\frac{5}{8} - 1\frac{1}{4}$ _____

6. $5\frac{3}{4} - 2\frac{1}{8}$ _____

7. $4\frac{2}{3} - 1\frac{1}{2}$ _____

8. $3\frac{1}{3} - 2\frac{1}{4}$ _____

9. $6\frac{2}{5} - 3\frac{1}{3}$ _____

10. $4\frac{5}{6} - 3\frac{3}{4}$ _____

Solving Problems

Solve. Reduce your answers, if necessary.

11. The restaurant where Zahra works has $3\frac{1}{2}$ apple pies. She serves $1\frac{1}{4}$ pies. How much pie is left?

12. There are $3\frac{2}{3}$ hours until Tom's favorite TV program. He runs errands for $1\frac{1}{2}$ hours. How much time is left until Tom's program is on?

_____

_____

Check your answers on page 79.

Solving Problems

16

Subtracting Like Mixed Numbers With Borrowing

Pon Chinn has $2\frac{1}{4}$ cups of cooking oil. He used $1\frac{3}{4}$ cups in a recipe. He wants to figure out how much oil he has left. Chinn will be subtracting mixed numbers with like fractions. The problem is in subtracting the fraction portions. How can Chinn subtract $\frac{3}{4}$ cup from $\frac{1}{4}$ cup? How much oil does he have left?

Think

When the second fraction is larger than the first fraction, you must *borrow* from the whole number before subtracting the mixed numbers.

When you subtract whole numbers, you always borrow 10 from the column on the left and add it to the column you were subtracting. Now you need to borrow from the ones column so that you can add to the fraction column.

You will be borrowing 1 whole, writing the 1 as a fraction, and adding it to the fraction column.

To write a fraction equal to 1, use the same number as both the numerator and the denominator. $\frac{2}{2}$, $\frac{3}{3}$, and $\frac{4}{4}$ are examples of improper fractions equal to 1.

To subtract mixed numbers with like fractions when you have to borrow:

Rewrite the mixed number by taking 1 from the whole number portion. Write the 1 as a like improper fraction with the same number as the numerator and the denominator. Add the like fraction to the fraction you are subtracting from.

Subtract the mixed numbers with like fractions.

You can use the same method to subtract a mixed fraction from a whole number alone. Borrow 1 from the whole number and write the whole number as a like improper fraction.

Do

$2\frac{1}{4} - 1\frac{3}{4}$

Step 1. Borrow. Express the 1 you borrow as $\frac{4}{4}$ and add it to $\frac{1}{4}$ in the fractions column.

$$\overset{1}{\cancel{2}}\frac{1}{4} + \frac{4}{4} \qquad 1\frac{5}{4}$$
$$-\ 1\frac{\mathbf{5}}{4} \qquad\quad -\ 1\frac{3}{4}$$

Step 2. Subtract the whole number parts.

$1 - 1 = 0$

Step 3. Subtract the fraction parts.

$\dfrac{5}{4} - \dfrac{3}{4} = \dfrac{2}{4}$

Step 4. Reduce the fraction to lowest terms.

$\dfrac{2}{4} = \dfrac{1}{2}$

Chinn has $\frac{1}{2}$ cup of cooking oil left.

Try These

Subtract these mixed numbers. Reduce the answer if necessary.

1. $3\frac{1}{3} - 1\frac{2}{3}$

   Rewrite the first mixed number by borrowing.

$$\overset{2}{\cancel{3}}\frac{1}{3} + \frac{3}{3} \qquad 2\frac{4}{3}$$
$$-\ 1\frac{2}{3} \qquad\quad -\ 1\frac{2}{3}$$

   Subtract the whole number parts.

   $2 - 1 = 1$

   Subtract the fraction parts.

   $\dfrac{4}{3} - \dfrac{2}{3} = \dfrac{?}{3}$

   The answer is $1\frac{?}{3}$. _____

   Subtract this mixed number from a whole number.

2. $5 - 2\frac{3}{8}$

   Borrow and write the whole number as a mixed number.

$$\overset{4}{\cancel{5}} \qquad\qquad 4\frac{8}{8}$$
$$-\ 2\frac{3}{8} \qquad\quad -\ 2\frac{3}{8}$$

   Subtract the whole number parts.

   $4 - 2 = ?$

   Subtract the fraction parts.

   $\dfrac{8}{8} - \dfrac{3}{8} = \dfrac{?}{8}$

   The answer is $?\,\dfrac{?}{8}$.

Subtract. Reduce your answers to lowest terms.

3. $2\frac{3}{5} - 1\frac{4}{5}$ _____

4. $3 - \frac{3}{4}$ _____

5. $5\frac{1}{8} - 3\frac{2}{8}$ _____

6. $4\frac{1}{6} - 2\frac{5}{6}$ _____

7. $8 - 3\frac{3}{5}$ _____

8. $5 - 1\frac{5}{6}$ _____

9. $3\frac{1}{3} - 1\frac{2}{3}$ _____

10. $9 - 5\frac{1}{2}$ _____

Solving Problems

Solve. Reduce answers to lowest terms.

11. Nick has $5\frac{5}{8}$ feet of wood. He cuts off a piece $2\frac{7}{8}$ feet long. How much wood does Nick have left?

12. Mia's cat weighs 8 pounds before her diet. She loses $\frac{3}{4}$ of a pound. How much does the cat weigh now?

_____   _____

Check your answers on pages 79–80.

17

Subtracting Unlike Mixed Numbers With Borrowing

Juice is on sale at the grocery store. Mary buys a $2\frac{1}{6}$ gallon bottle. She pours $1\frac{1}{3}$ gallons in a container for her sister. Mary wants to know how much juice she has left.

Think

Mary knows she has to subtract these mixed numbers. She knows how to subtract like mixed numbers with borrowing. These mixed numbers, however, do not have like fractions. Before she can subtract, Mary has to change the fractions to like fractions.

To subtract unlike mixed numbers with borrowing:

Find a common denominator and raise the terms of the fraction parts as needed.

Subtract the like mixed numbers with borrowing.

Do

$$2\frac{1}{6} - 1\frac{1}{3}$$

Step 1. Raise the terms of the second fraction to the least common denominator (6).

$$\frac{1}{3} \times \frac{2}{2} = \frac{2}{6} \qquad \begin{array}{r} 2\frac{1}{6} \\ -\ 1\frac{2}{6} \\ \hline \end{array}$$

Step 2. Borrow and rewrite the first mixed number.

$$\begin{array}{r} \overset{1}{2}\frac{1}{6} + \frac{6}{6} \\ -\ 1\frac{2}{6} \\ \hline \end{array} \qquad \begin{array}{r} 1\frac{7}{6} \\ -\ 1\frac{2}{6} \\ \hline \end{array}$$

Step 3. Subtract the whole number parts. $\qquad 1 - 1 = 0$

Step 4. Subtract the fraction parts. $\qquad \dfrac{7}{6} - \dfrac{2}{6} = \dfrac{5}{6}$

Mary has $\frac{5}{6}$ of a gallon of juice left in the original bottle.

Try These

Subtract these unlike mixed numbers with borrowing. Reduce to lowest terms, if necessary.

1. $3\frac{1}{2} - 1\frac{3}{4}$

Raise the terms of the first fraction to the least common denominator.

$$\frac{1}{2} \times \frac{2}{2} = \frac{2}{4} \qquad 3\frac{2}{4}$$
$$-1\frac{3}{4}$$

Borrow and rewrite the first mixed number.

$$\overset{2}{\cancel{3}}\frac{2}{4} + \frac{4}{4} \qquad 2\frac{6}{4}$$
$$-1\frac{3}{4} \qquad\qquad -1\frac{3}{4}$$

Subtract the whole number parts.

$$2 - 1 = 1$$

Subtract the fraction parts.

$$\frac{6}{4} - \frac{3}{4} = \frac{?}{4}.$$

The answer is $1\frac{?}{4}$. _____

2. $4\frac{1}{4} - 2\frac{1}{3}$

Raise the terms of the fractions to the least common denominator (12).

$$\frac{1}{4} \times \frac{3}{3} = \frac{3}{12} \qquad \frac{1}{3} \times \frac{4}{4} = \frac{4}{12}$$

$$4\frac{3}{12} \qquad 2\frac{4}{12}$$

Borrow and rewrite the first mixed number.

$$\overset{3}{\cancel{4}}\frac{3}{12} + \frac{12}{12} \qquad 3\frac{15}{12}$$
$$-2\frac{4}{12} \qquad\qquad -2\frac{4}{12}$$

Subtract the whole number parts.

$$3 - 2 = 1$$

Subtract the fraction parts.

$$\frac{?}{12} - \frac{4}{12} = \frac{?}{12}$$

The answer is $1\frac{?}{12}$. _____

Subtract. Reduce to lowest terms, if necessary.

3. $3\frac{1}{4} - 1\frac{1}{2}$ _____

4. $4\frac{3}{8} - 1\frac{3}{4}$ _____

5. $3\frac{2}{3} - 1\frac{5}{6}$ _____

6. $4\frac{1}{4} - 2\frac{3}{8}$ _____

7. $5\frac{1}{3} - 2\frac{3}{4}$ _____

8. $5\frac{1}{6} - 1\frac{3}{4}$ _____

9. $6\frac{3}{10} - 2\frac{3}{5}$ _____

10. $8\frac{2}{5} - 5\frac{3}{4}$ _____

Solving Problems

Solve. Reduce your answers to lowest terms.

11. Emil has to use $1\frac{3}{4}$ pounds of potatoes to cook dinner. He has a $5\frac{1}{2}$ pound bag of potatoes. How many pounds of potatoes will Emil have left in the bag after cooking dinner?

12. Denise fills a $12\frac{1}{5}$ ounce bottle with $8\frac{1}{3}$ ounces of baby formula. How many more ounces can the bottle hold?

_____   _____

Check your answers on page 80.

18

Review of Addition and Subtraction

Add. Reduce your answer to lowest terms.

1. $\dfrac{1}{8} + \dfrac{3}{8}$ _____

2. $\dfrac{2}{5} + \dfrac{1}{5}$ _____

3. $\dfrac{2}{9} + \dfrac{4}{9}$ _____

4. $\dfrac{2}{3} + \dfrac{1}{6}$ _____

5. $\dfrac{3}{8} + \dfrac{1}{4}$ _____

6. $\dfrac{2}{9} + \dfrac{2}{3}$ _____

7. $2\dfrac{1}{3} + 1\dfrac{2}{3}$ _____

8. $3\dfrac{1}{5} + 2\dfrac{2}{5}$ _____

9. $3\dfrac{1}{4} + 1\dfrac{1}{2}$ _____

10. $2\dfrac{3}{4} + 2\dfrac{2}{3}$ _____

Subtract. Reduce your answer to lowest terms.

11. $\dfrac{5}{6} - \dfrac{1}{6}$ _____

12. $\dfrac{3}{4} - \dfrac{1}{4}$ _____

13. $\dfrac{7}{8} - \dfrac{3}{8}$ _____

14. $\dfrac{5}{8} - \dfrac{1}{4}$ _____

15. $\dfrac{5}{6} - \dfrac{2}{3}$ _____

16. $3\dfrac{3}{4} - 1\dfrac{1}{4}$ _____

17. $5\dfrac{7}{8} - 2\dfrac{3}{8}$ _____

18. $4\dfrac{2}{3} - 1\dfrac{1}{6}$ _____

19. $3\dfrac{3}{4} - 2\dfrac{2}{3}$ _____

20. $6\dfrac{1}{4} - 4\dfrac{3}{4}$ _____

21. $8\dfrac{1}{6} - 4\dfrac{5}{6}$ _____

22. $4\dfrac{1}{4} - 1\dfrac{1}{2}$ _____

23. $6\dfrac{1}{6} - 3\dfrac{1}{4}$ _____

Solving Problems

Solve. Reduce your answer to lowest terms.

24. Barry ran $2\frac{2}{3}$ miles one day. The next day he ran $3\frac{3}{4}$ miles. How many miles did Barry run together?

_____

25. Andrea had $5\frac{1}{2}$ pounds of flour. She used $\frac{3}{4}$ pound baking bread. How much flour did Andrea have left?

_____

26. Omri is studying for a math test. He studied $1\frac{1}{2}$ hours today. Yesterday he studied $2\frac{3}{4}$ hours. How many hours has he studied for the test?

_____

27. Maria is weighing packages so that she can put the right amount of postage on them. A small package weighs $8\frac{1}{2}$ ounces. A larger package weighs 12 ounces. How much more does the larger package weigh than the smaller one?

_____

28. Joon Bok bought $9\frac{1}{4}$ yards of canvas to make two banners. He uses $5\frac{7}{8}$ yards to make the first banner. How much canvas does he have left to make the second banner?

_____

29. A dentist is scheduling a patient for a surface filling and a crown. If a surface filling usually takes $\frac{1}{3}$ hour to complete and a crown takes $1\frac{1}{4}$ hour, how much time will the dentist need to set aside for the patient?

_____

Check your answers on page 80–81.

Finding Fractional Parts

Jonas will be taking a math test at school. There are 15 questions on the test. The teacher says that $\frac{1}{3}$ of the questions are multiple-choice questions. How many multiple-choice questions will Jonas find on the test?

Think

Jonas needs to find $\frac{1}{3}$ of 15. The word "of" following a fraction means you must multiply the fraction by the number. Jonas needs to solve the problem: $\frac{1}{3} \times 15$.

To multiply a fraction by a whole number, multiply the numerator by the whole number. Put the result over the denominator of the fraction. If the answer is an improper fraction, change it to a mixed number. Always reduce the fraction part of the answer to lowest terms.

Do

$\frac{1}{3} \times 15$

Step 1. Multiply the numerator of the fraction by the whole number.

$1 \times 15 = 15$

Step 2. Put the result over the denominator of the fraction.

$\frac{15}{3}$

Step 3. Turn the improper fraction into a mixed number. Divide the numerator by the denominator.

$15 \div 3 = 5$

Jonas finds that there will be five multiple choice questions on the test.

Try These

Find the fractional part of the whole number. Change improper fractions to mixed numbers. Reduce to lowest terms.

1. $\frac{3}{4} \times 11$ $\qquad$ $3 \times 11 = 33$

$\frac{33}{4}$ $\qquad$ $33 \div 4 = ?\frac{?}{?}$ _____

47

2. $\dfrac{3}{8} \times 12$

$3 \times 12 = 36$

$\dfrac{36}{8}$ $36 \div 8 = 4\dfrac{?}{8}$ which reduces to $4\dfrac{?}{?}$ _____

PRACTICE

Find the fractional part of the whole number. Change improper fractions to mixed numbers. Reduce to lowest terms.

3. $\dfrac{1}{4} \times 8$ _____

4. $\dfrac{1}{3} \times 7$ _____

5. $\dfrac{2}{3} \times 9$ _____

6. $\dfrac{2}{3} \times 11$ _____

7. $\dfrac{1}{6} \times 15$ _____

8. $\dfrac{2}{5} \times 18$ _____

9. $\dfrac{7}{8} \times 3$ _____

10. $\dfrac{5}{6} \times 10$ _____

Solving Problems

Solve. Reduce to lowest terms.

11. Brad applies for a new job. He is told he will work for 8 hours a day. He also finds out that he will spend $\frac{1}{4}$ of the time making deliveries. How many hours a day will he be making deliveries?

12. Rose plays softball for her school. Early in the season she has 25 hits. $\frac{2}{5}$ of her hits are doubles. How many doubles did Rose hit?

_____   _____

Check your answers on page 81.

Multiplying Proper Fractions

One-third of the dresses in stock at a boutique are blue. Of those, $\frac{1}{4}$ are a solid color. The manager wants to find the fraction of the dresses in stock that are blue and solid.

Think

The manager needs to find a part of a fraction. That is, she needs to find $\frac{1}{4}$ of $\frac{1}{3}$. This is a multiplication problem with two proper fractions. The result will be a proper fraction that is smaller than either of the two fractions that are multiplied.

To multiply proper fractions:

Multiply the numerators to find the numerator of the answer.

Multiply the denominators to find the denominator of the answer.

Reduce the result to lowest terms, if necessary.

Do

$$\frac{1}{4} \times \frac{1}{3} =$$

Step 1. Multiply the numerators. $1 \times 1 = 1$

Step 2. Multiply the denominators. $4 \times 3 = 12$
Of the dresses in stock, $\frac{1}{12}$ are blue and solid.

Try These

Multiply these proper fractions. Reduce the answer, if necessary.

1. $\frac{1}{2} \times \frac{3}{4}$

Multiply the numerators. $1 \times 3 = 3$

Multiply the denominators. $2 \times 4 = 8$

The answer is $\frac{?}{?}$. _____

2. $\dfrac{2}{3} \times \dfrac{1}{2}$

Multiply the numerators. $\qquad\qquad 2 \times 1 = ?$

Multiply the denominators. $\qquad\qquad 3 \times ? = ?$

The answer is $\dfrac{?}{?}$.

Reduced to lowest terms, the answer is $\dfrac{?}{?}$. _____

PRACTICE

Multiply. Reduce the answer, if necessary.

3. $\dfrac{1}{2} \times \dfrac{1}{4}$ _____   4. $\dfrac{3}{4} \times \dfrac{1}{3}$ _____

5. $\dfrac{2}{3} \times \dfrac{3}{4}$ _____   6. $\dfrac{1}{4} \times \dfrac{1}{5}$ _____

7. $\dfrac{3}{5} \times \dfrac{2}{3}$ _____   8. $\dfrac{3}{4} \times \dfrac{2}{5}$ _____

9. $\dfrac{1}{6} \times \dfrac{2}{3}$ _____   10. $\dfrac{5}{6} \times \dfrac{3}{4}$ _____

Solving Problems

Multiply. Reduce the answer, if necessary.

11. Kwang-Son buys $\frac{1}{2}$ loaf of bread. He gives $\frac{3}{4}$ to his brother. What fraction of a loaf of bread is $\frac{3}{4}$ of $\frac{1}{2}$?

12. Katrina traveled $\frac{4}{5}$ of a mile. She ran $\frac{2}{3}$ of the way. What fraction of a mile did Katrina run?

_____   _____

Check your answers on page 82.

21

Canceling

Susannah's boss tells her that $\frac{5}{24}$ of the company's employees need to sign up for their new health insurance. Of those, $\frac{2}{5}$ also need to sign up for retirement benefits. Her boss asks her to figure out what fraction of the employees need to sign up for both benefits.

Susannah knows she can solve the problem by multiplying $\frac{5}{24} \times \frac{2}{5}$, but she wishes she didn't have to work with such large numbers.

Is there a way to simplify the multiplication of these proper fractions?

Think

Before multiplying fractions, try and find a whole number that divides evenly into the numerator of one fraction and the denominator of the other. The process is similar to the one used to reduce fractions to lowest terms.

Canceling means to simplify a problem by dividing the numerator and denominator by the same whole number. The resulting lower terms are easier to multiply. Canceling can eliminate the need to reduce to lowest terms.

To cancel fractions before multiplying them:

Look at the numerator of the first fraction and the denominator of the second fraction.

Divide them by the largest whole number you can find.

Repeat the steps for the denominator of the first fraction and the numerator of the second fraction.

Do

$$\frac{5}{24} \times \frac{2}{5}$$

Step 1. Find a whole number that divides into 5 and 5 evenly.   5

Divide the numerator and denominator by 5.
Cross out the 5's and write 1's to the side.

$$(5 \div 5 = 1) \quad \frac{\cancel{5}^{1}}{24} \times \frac{2}{\cancel{5}_{1}} \quad (5 \div 5 = 1)$$

Step 2. Find a whole number that divides into 24 and 2 evenly.   2

Divide the numerator and denominator by 2.
Cross out the numbers and write in the new
numerator and denominator.

$$(24 \div 2 = 12) \quad \frac{\cancel{5}^{1}}{\cancel{24}_{12}} \times \frac{\cancel{2}^{1}}{\cancel{5}_{1}} \quad (2 \div 2 = 1)$$

Step 3. Multiply the canceled fractions.

$$\frac{\cancel{5}^{1}}{\cancel{24}_{12}} \times \frac{\cancel{2}^{1}}{\cancel{5}_{1}} = \frac{1}{12}$$

The answer is $\frac{1}{12}$. One-twelfth of the employees
need to sign up for both benefits.

Try These

Cancel these fractions before multiplying them.

1. $\dfrac{3}{4} \times \dfrac{2}{3}$ _____

$$\frac{\cancel{3}^{1}}{\cancel{4}_{?}} \times \frac{\cancel{2}^{?}}{\cancel{3}_{1}}$$

2. $\dfrac{3}{8} \times \dfrac{5}{9}$ _____

$$\frac{3}{8} \times \frac{5}{9} = \text{_____}$$

Cancel these fractions before multiplying them.

3. $\dfrac{3}{4} \times \dfrac{2}{3}$ _____

4. $\dfrac{3}{4} \times \dfrac{1}{6}$ _____

5. $\dfrac{1}{4} \times \dfrac{4}{5}$ _____

6. $\dfrac{3}{8} \times \dfrac{2}{3}$ _____

7. $\dfrac{5}{9} \times \dfrac{3}{10}$ _____

8. $\dfrac{5}{8} \times \dfrac{4}{5}$ _____

9. $\dfrac{7}{10} \times \dfrac{5}{7}$ _____

10. $\dfrac{8}{9} \times \dfrac{3}{8}$ _____

Solving Problems

Cancel these fractions before multiplying them.

11. Martha had $\frac{3}{8}$ of a pumpkin pie left after a Thanksgiving dinner. She split the pie into thirds and gave one piece to each of her children. How much is $\frac{1}{3}$ of $\frac{3}{8}$ of a pie?

_____

12. Teresa has a bag of rice with $\frac{5}{6}$ of a pound of rice left. She wants to use $\frac{2}{3}$ of the rice. How many pounds of rice does she use?

_____

Check your answers on page 82.

22

Multiplying Mixed Numbers

Roger is making potato salad. The recipe he is using calls for $3\frac{1}{3}$ cups of diced potatoes. He needs to make $2\frac{1}{2}$ times the amount of the recipe. Roger needs to figure out how many cups of potatoes to dice.

Think

Roger must multiply $3\frac{1}{3} \times 2\frac{1}{2}$. He needs to multiply mixed numbers.

To multiply mixed numbers:

Change the numbers to improper fractions.

Cancel, if possible.

Multiply the fractions.

Turn the result into a mixed number.

Do

$3\frac{1}{3} \times 2\frac{1}{2}$

Step 1. Turn the numbers into improper fractions.

$$3\frac{1}{3} = 3 \times 3 + 1 = \frac{10}{3}$$

$$2\frac{1}{2} = 2 \times 2 + 1 = \frac{5}{2}$$

Step 2. Cancel.

$$\frac{\overset{5}{\cancel{10}}}{3} \times \frac{5}{\underset{1}{\cancel{2}}} =$$

Step 3. Multiply the fractions.

$$\frac{\overset{5}{\cancel{10}}}{3} \times \frac{5}{\underset{1}{\cancel{2}}} = \frac{25}{3}$$

Step 4. Change the result into a mixed number.

$$25 \div 3 = 8\frac{1}{3}$$

Roger needs to dice $8\frac{1}{3}$ cups of potatoes.

Try These

Multiply these mixed numbers.

1. $1\frac{1}{2} \times 2\frac{1}{2}$

Turn the mixed numbers into improper fractions.

$2 \times 1 + 1 = 3 \quad \frac{3}{2} \qquad 2 \times 2 + 1 = 5 \quad \frac{5}{2}$

It isn't possible to cancel. Multiply the fractions. $\quad \frac{3}{2} \times \frac{5}{2} = \frac{15}{4}$

Change the answer to a mixed number. $\qquad 15 \div 4 = 3\frac{?}{?} \; \underline{\quad\quad}$

2. $2\frac{1}{4} \times 1\frac{5}{9}$

Turn the mixed numbers into improper fractions.

$4 \times 2 + 1 = 9 \quad \frac{9}{4} \qquad 9 \times 1 + 5 = 14 \quad \frac{14}{9}$

Cancel and multiply. $\qquad \overset{1}{\cancel{9}} \times \overset{7}{\cancel{14}} = \frac{?}{?}$
$\qquad\qquad\qquad\quad \underset{2}{\cancel{4}} \quad \underset{1}{\cancel{9}}$

Write the answer as a mixed number. $\qquad$ The answer is $3\frac{?}{?}.$ $\underline{\quad\quad}$

--- **PRACTICE** ---

Multiply.

3. $2\frac{2}{3} \times 1\frac{1}{4}$ $\underline{\quad\quad}$ $\qquad\qquad$ **4.** $1\frac{1}{2} \times 1\frac{1}{2}$ $\underline{\quad\quad}$

5. $1\frac{1}{4} \times 2\frac{1}{3}$ $\underline{\quad\quad}$ $\qquad\qquad$ **6.** $1\frac{1}{8} \times 1\frac{2}{3}$ $\underline{\quad\quad}$

7. $1\frac{1}{6} \times 1\frac{1}{2}$ $\underline{\quad\quad}$ $\qquad\qquad$ **8.** $2\frac{1}{4} \times 1\frac{3}{7}$ $\underline{\quad\quad}$

9. $1\frac{5}{6} \times 2\frac{1}{4}$ $\underline{\quad\quad}$ $\qquad\qquad$ **10.** $2\frac{2}{9} \times 1\frac{7}{8}$ $\underline{\quad\quad}$

Multiply.

11. Tina has $3\frac{1}{2}$ feet of rope. She needs $2\frac{1}{7}$ times that amount of rope. How many feet of rope does Tina need?

12. Roberto uses $2\frac{1}{2}$ gallon of gasoline to drive to his aunt's house. How many gallons of gasoline will he use if he travels $3\frac{3}{5}$ times as far?

_____

_____

Check your answers on page 82.

23

The Relationship Between Multiplication and Division

Marty has 12 ounces of hamburger meat. He wants to divide the meat to make three hamburgers. Marty has to figure out how many ounces each hamburger patty will weigh.

Think

Marty can divide 12 ounces by 3 to find that each hamburger will weigh 4 ounces.

$$12 \div 3 = 4$$

But with what Marty knows about fractions, he can also multiply 12 by $\frac{1}{3}$ to get the same answer.

$$12 \times \frac{1}{3} = \frac{12}{3} = 4$$

Marty sees that dividing a number gives him the same result as multiplying the number by its *reciprocal*. The reciprocal of a number is found by inverting the numerator and denominator of the fraction. For example, the reciprocal of $\frac{3}{4}$ is $\frac{4}{3}$. To divide a number by a fraction, invert the fraction and multiply.

A whole number can be written as a fraction by using the whole number as the numerator and using 1 as the denominator. 3 is the same as the improper fraction $\frac{3}{1}$. The reciprocal of $\frac{3}{1}$ is $\frac{1}{3}$.

Do

Before dividing a number, invert the divisor (3) into its reciprocal.

Step 1. Turn the whole number into a fraction. $3 = \frac{3}{1}$

Step 2. Invert the fraction. $\frac{3}{1}$ inverts to $\frac{1}{3}$

Try These

Invert these numbers to find their reciprocal.

1. $\frac{3}{4}$ Invert the numerator and denominator.

             $\frac{4}{?}$ is the reciprocal of $\frac{3}{4}$. _____

2. 6   Write the whole number as a fraction. $\frac{6}{?}$

             Invert the numerator and denominator.

             $\frac{?}{?}$ is the reciprocal of 6. _____

PRACTICE

Write the reciprocals of these numbers.

3. $\frac{1}{2}$ _____   4. $\frac{3}{8}$ _____

5. $\frac{9}{10}$ _____   6. $\frac{2}{3}$ _____

7. $\frac{8}{5}$ _____   8. $\frac{9}{4}$ _____

9. 4 _____   10. 9 _____

Solving Problems

Solve.

11. Bill wants to divide a length of wood into $\frac{2}{3}$ foot pieces. He knows he needs to multiply the length of the wood by the reciprocal of $\frac{2}{3}$. What is the reciprocal?

12. Amanda has a pitcher of lemonade. She pours $\frac{3}{4}$ of a cup of lemonade into each glass. To find out the number of glasses she can pour from the pitcher, she needs to divide by $\frac{3}{4}$ or multiply by its reciprocal. What is the reciprocal of $\frac{3}{4}$?

_____   _____

Check your answers on page 83.

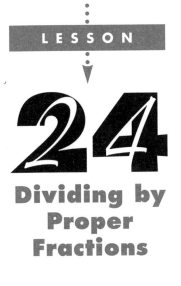

24

Dividing by Proper Fractions

Mitsue has 6 pounds of potting soil. She needs $\frac{3}{4}$ of a pound for each pot. Mitsue needs to know how many pots she can fill.

Think

Mitsue has to divide 6 by $\frac{3}{4}$. To divide by a fraction, she must invert the fraction to its reciprocal and then multiply by the reciprocal. Mitsue also remembers that a whole number can be turned into a fraction by writing the whole number as the numerator and 1 as the denominator.

To divide a number by a fraction:

Invert the fraction to its reciprocal.

Cancel, if possible.

Multiply the fractions.

If the result is an improper fraction, turn it into a mixed number.

Do

$6 \div \dfrac{3}{4}$

Step 1. Invert the fraction to its reciprocal. $\dfrac{3}{4}$ inverts to $\dfrac{4}{3}$.

Step 2. Write the whole number 6 as an improper fraction. $6 = \dfrac{6}{1}$

Step 3. Cancel the fractions. $\dfrac{\overset{2}{\cancel{6}}}{1} \times \dfrac{4}{\underset{1}{\cancel{3}}} =$

Step 4. Multiply. $\dfrac{\overset{2}{\cancel{6}}}{1} \times \dfrac{4}{\underset{1}{\cancel{3}}} = \dfrac{8}{1}$

$\frac{8}{1} = 8$. Mitsue will be able to fill 8 pots.

59

Try These

Divide.

1. $10 \div \dfrac{4}{5}$

   Invert the fraction to its reciprocal.

   $\dfrac{4}{5}$ inverts to $\dfrac{5}{4}$

   Write the whole number as an improper fraction.

   $10 = \dfrac{10}{1}$

   Cancel and multiply.

   $\dfrac{10}{1} \times \dfrac{5}{4}$ $\dfrac{\overset{5}{\cancel{10}}}{1} \times \dfrac{5}{\underset{2}{\cancel{4}}} = \dfrac{25}{2}$

   Change the improper fraction to a mixed number.

   $\dfrac{25}{2} = ?\dfrac{?}{?}$ _____

2. $\dfrac{3}{4} \div \dfrac{2}{5}$

   Invert the fraction to its reciprocal.

   $\dfrac{2}{5}$ inverts to $\dfrac{5}{2}$ $\dfrac{3}{4} \times \dfrac{5}{2} = \dfrac{?}{?}$

   Change the improper fraction to a mixed number.

   $?\dfrac{?}{?}$ _____

PRACTICE

Divide.

3. $\dfrac{1}{2} \div \dfrac{1}{4}$ _____

4. $\dfrac{1}{3} \div \dfrac{5}{6}$ _____

5. $\dfrac{2}{3} \div \dfrac{3}{4}$ _____

6. $\dfrac{3}{4} \div \dfrac{2}{3}$ _____

7. $18 \div \dfrac{3}{4}$ _____

8. $15 \div \dfrac{5}{8}$ _____

9. $8 \div \dfrac{1}{5}$ _____

10. $10 \div \dfrac{2}{3}$ _____

Solving Problems

Divide.

11. Willis loads newspapers into machines on the street. In a busy downtown area, there are newspaper machines every $\frac{2}{3}$ of a block. Willis's route covers 8 city blocks. How many stops does he make to put papers in machines?

12. At an animal care facility, Soloman pours $\frac{5}{6}$ of a gallon of milk into $\frac{3}{8}$ gallon bowls. The bowls are put into the animals' cages. How many bowls will Soloman be able to fill?

_____

_____

Check your answers on page 83.

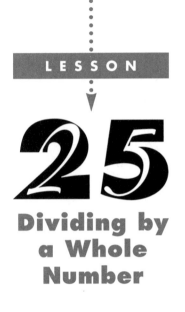

Juan and a friend agree to oversee the gardening work at a large apartment complex. Juan agrees to do $\frac{7}{8}$ of the work. Then Juan hires 6 workers and divides his work equally among them. What fraction of Juan's work will each of the workers complete?

Think

You need to divide $\frac{7}{8}$ by 6. To divide a fraction, you need to invert the number you are dividing by and multiply. Since 6 is a whole number, you will need to write it as an improper fraction and then invert it.

To divide a fraction by a whole number:

Write the whole number as an improper fraction.

Invert the improper fraction.

Cancel, if possible.

Multiply the fractions.

Do

$$\frac{7}{8} \div 6$$

Step 1. Write the whole number as an improper fraction and invert it.

$\frac{1}{6}$ is the reciprocal of $\frac{6}{1}$.

$$6 = \frac{6}{1}$$

Step 2. Multiply the fractions.

$$\frac{7}{8} \times \frac{1}{6} = \frac{7}{48}$$

Each worker will do $\frac{7}{48}$ of Juan's work.

LESSON

25

Dividing by a Whole Number

Try These

Divide each of these fractions by the whole number.

1. $\dfrac{1}{4} \div 2$

    Write an improper fraction and invert it.

    $\dfrac{1}{2}$ is the reciprocal of $\dfrac{2}{1}$.

    Multiply the fractions.

    $2 = \dfrac{2}{1}$

    $\dfrac{1}{4} \times \dfrac{1}{2} = \dfrac{?}{?}$ _____

2. $\dfrac{5}{6} \div 3$

    Write an improper fraction and invert it.

    The reciprocal of $\dfrac{3}{1}$ is $\dfrac{?}{?}$.

    Multiply the fractions.

    $3 = \dfrac{3}{1}$

    $\dfrac{5}{6} \times \dfrac{?}{?} = \dfrac{?}{?}.$ _____

PRACTICE

Divide.

3. $\dfrac{1}{2} \div 4$ _____

4. $\dfrac{2}{3} \div 4$ _____

5. $\dfrac{1}{6} \div 2$ _____

6. $\dfrac{3}{4} \div 6$ _____

7. $\dfrac{5}{6} \div 5$ _____

8. $\dfrac{2}{3} \div 8$ _____

9. $\dfrac{4}{5} \div 3$ _____

10. $\dfrac{5}{8} \div 10$ _____

Solving Problems

Solve.

11. Manuel needs to divide $\frac{3}{4}$ of a cup of liquid fertilizer into 6 equal parts. How much will each part contain?

    _____

12. Joanne bought $\frac{3}{4}$ pound of chopped meat. She makes 15 meatballs of the same size. How much does each meatball weigh?

    _____

Check your answers on pages 83–84.

26

Dividing Mixed Numbers

Chris has $8\frac{1}{4}$ pounds of oranges to divide into $\frac{3}{4}$ pound bags. How many bags of oranges he can fill?

Think

Chris needs to divide $8\frac{1}{4}$ by $\frac{3}{4}$. But how can you divide a mixed number by a fraction? First, change the mixed number to an improper fraction. Then you can invert and multiply. Remember to use canceling to make the work easier. If the answer is an improper fraction, change it to a mixed number and reduce to lowest terms.

To divide a mixed number:

Change the mixed number into an improper fraction.

Invert the fraction you are dividing by.

Cancel if possible.

Multiply the fractions.

Simplify the answer. Change improper fractions to mixed numbers and reduce to lowest terms.

Do

$$8\frac{1}{4} \div \frac{3}{4}$$

Step 1. Turn the mixed number into an improper fraction.

$$4 \times 8 + 1 = 33 \qquad \frac{33}{4}$$

Step 2. Invert the divisor into its reciprocal.

$\frac{4}{3}$ is the reciprocal of $\frac{3}{4}$

Step 3. Cancel.

$$\frac{\overset{11}{\cancel{33}}}{\underset{1}{\cancel{4}}} \times \frac{\overset{1}{\cancel{4}}}{\underset{1}{\cancel{3}}} =$$

Step 4. Multiply the fractions.

$$\frac{11}{1} \times \frac{1}{1} = \frac{11}{1} = 11$$

Chris will be able to fill 11 bags with oranges.

Try These

Divide.

1. $2\frac{2}{3} \div 6$

   Write the mixed number as an improper fraction.

   $3 \times 2 + 2 = 8 \qquad \dfrac{8}{3}$

   Write the reciprocal of the number you are dividing by.

   $\frac{1}{6}$ is the reciprocal of 6.

   Cancel.

   $\dfrac{\overset{4}{\cancel{8}}}{3} \times \dfrac{1}{\underset{3}{\cancel{6}}}$

   Multiply the fractions.

   $\dfrac{4}{3} \times \dfrac{1}{3} = \dfrac{?}{?}$ _____

2. $3\frac{3}{4} \div \frac{2}{3}$

   Write the mixed number as an improper fraction.

   $4 \times 3 + 3 = 15 \qquad \dfrac{15}{4}$

   Invert the number you are dividing by.

   $\frac{3}{2}$ is the reciprocal of $\frac{2}{3}$

   Multiply the fractions.

   $\dfrac{15}{4} \times \dfrac{3}{2} = \dfrac{?}{?}$ _____

   Write the improper fraction as a mixed number.

   $\dfrac{?}{?} = ?\dfrac{?}{?}$ _____

PRACTICE

Divide these mixed numbers.

3. $1\frac{1}{4} \div \frac{1}{2}$ _____

4. $2\frac{2}{3} \div 4$ _____

5. $1\frac{1}{3} \div \frac{3}{4}$ _____

6. $2\frac{2}{5} \div 6$ _____

7. $3\frac{3}{4} \div \frac{1}{5}$ _____

8. $5\frac{1}{2} \div \frac{4}{5}$ _____

9. $1\frac{5}{6} \div \frac{1}{2}$ _____

10. $2\frac{4}{5} \div \frac{3}{10}$ _____

Solving Problems

Divide.

11. Ivan ran $4\frac{1}{2}$ miles in $\frac{3}{4}$ of a hour. How many miles did Ivan run each hour? [HINT: Divide $4\frac{1}{2}$ by $\frac{3}{4}$ to find miles per hour.]

12. Gigi buys a $10\frac{1}{8}$ ounce can of baked beans. She plans to use the beans in a recipe that serves 9 people. If the beans are divided equally among the servings, how many ounces of baked beans will be in each serving?

_____ _____

Check your answers on page 84.

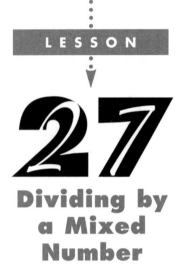

Marco has a room that is 18 feet wide. He needs to mark off $1\frac{1}{5}$ foot lengths. Marco has to divide 18 by the mixed number $1\frac{1}{5}$. How can he divide by a mixed number?

Think

You know that to divide by a fraction, you must invert that fraction and multiply. When the number you are dividing by is a mixed number, turn it into an improper fraction first.

To divide by a mixed number:

Turn the mixed number into an improper fraction.

Invert the improper fraction and write a multiplication problem.

Cancel the fractions, if possible.

Multiply the fractions.

If the answer is an improper fraction, write it as a mixed number. Reduce to lowest terms if necessary.

Do

$18 \div 1\frac{1}{5}$

Step 1. Turn the whole number into a fraction.

$18 = \dfrac{18}{1}$

Step 2. Turn the mixed number into an improper fraction.

$5 \times 1 + 1 = 6 \qquad \dfrac{6}{5}$

Step 3. Write the reciprocal.

$\dfrac{5}{6}$ is the reciprocal of $\dfrac{6}{5}$

Step 4. Cancel.

$\dfrac{\overset{3}{\cancel{18}}}{1} \times \dfrac{5}{\underset{1}{\cancel{6}}} =$

Step 5. Multiply the fractions.

$\dfrac{3}{1} \times \dfrac{5}{1} = \dfrac{15}{1}$

Marco will mark off 15 sections.

Try These

Divide.

1. $\dfrac{3}{4} \div 1\dfrac{1}{2}$

Turn the mixed number into an improper fraction. $2 \times 1 + 1 = 3$ $\dfrac{3}{2}$

Invert the number you are dividing by.

$\dfrac{2}{3}$ is the reciprocal of $\dfrac{3}{2}$.

Cancel the fractions.

$$\dfrac{\overset{1}{\cancel{3}}}{\underset{2}{\cancel{4}}} \times \dfrac{\overset{1}{\cancel{2}}}{\underset{1}{\cancel{3}}} =$$

Multiply the fractions. $\dfrac{1}{2} \times \dfrac{1}{1} = \dfrac{?}{?}$ _____

2. $2\dfrac{2}{3} \div 1\dfrac{3}{5}$

Turn the mixed numbers into improper fractions. $3 \times 2 + 2 = 8$ $\dfrac{8}{3}$

$5 \times 1 + 3 = 8$ $\dfrac{8}{5}$

Invert the number you are dividing by.

$\dfrac{5}{8}$ is the reciprocal of $\dfrac{8}{5}$.

Cancel.

$$\dfrac{\overset{1}{\cancel{8}}}{3} \times \dfrac{5}{\underset{?}{\cancel{8}}} =$$

Multiply. $\dfrac{1}{3} \times \dfrac{?}{?} = \dfrac{?}{?}$

Write the answer as a mixed number. $\dfrac{?}{?} = ?\dfrac{?}{?}$ _____

Divide.

3. $3 \div 1\frac{1}{2}$ _____

4. $6 \div 1\frac{1}{3}$ _____

5. $10 \div 2\frac{4}{5}$ _____

6. $\frac{3}{4} \div 2\frac{1}{4}$ _____

7. $\frac{5}{6} \div 1\frac{2}{3}$ _____

8. $6\frac{2}{3} \div 2\frac{1}{2}$ _____

9. $7\frac{1}{3} \div 2\frac{3}{4}$ _____

10. $4\frac{4}{5} \div 2\frac{1}{2}$ _____

Solving Problems

Divide.

11. Sean works at a stable. He takes 12 apples from a bag and cuts them into smaller pieces. Then he gives $1\frac{1}{2}$ apples to each horse. How many horses does he give apples to?

_____

12. Jessica buys $12\frac{1}{2}$ cups of ice cream for her son's birthday party. She wants to give $1\frac{1}{4}$ cups of ice cream to each child. How many children can Jessica serve?

_____

Check your answers on pages 84–85.

Multiply. Reduce your answers to lowest terms.

1. $\dfrac{4}{5} \times \dfrac{1}{2}$ _____

2. $\dfrac{3}{4} \times \dfrac{2}{3}$ _____

3. $\dfrac{5}{7} \times \dfrac{7}{8}$ _____

4. $\dfrac{1}{4} \times \dfrac{8}{9}$ _____

5. $1\dfrac{1}{4} \times 2\dfrac{1}{3}$ _____

6. $3\dfrac{3}{5} \times 2\dfrac{2}{9}$ _____

7. $2\dfrac{1}{2} \times 2\dfrac{4}{5}$ _____

8. $4\dfrac{3}{8} \times 2\dfrac{4}{7}$ _____

Divide. Reduce your answers to lowest terms.

9. $\dfrac{3}{4} \div \dfrac{3}{5}$ _____

10. $\dfrac{5}{8} \div \dfrac{3}{4}$ _____

11. $\dfrac{2}{3} \div \dfrac{6}{7}$ _____

12. $8 \div \dfrac{3}{4}$ _____

13. $4 \div \dfrac{2}{3}$ _____

14. $\dfrac{5}{6} \div 2$ _____

15. $\dfrac{4}{5} \div 6$ _____

16. $2\dfrac{2}{3} \div \dfrac{1}{4}$ _____

17. $16 \div 2\dfrac{2}{3}$ _____

18. $\dfrac{5}{6} \div 1\dfrac{3}{7}$ _____

Solving Problems

Solve. Write improper fractions as mixed numbers and reduce to lowest terms.

19. Anna drove $10\frac{1}{4}$ miles using $1\frac{1}{2}$ gallon of gasoline. How many miles did she drive per gallon?

_____

20. Jean Paul uses a plant food that mixes $4\frac{1}{2}$ tablespoons of plant food with water. How much plant food should he use if he wants to make $2\frac{1}{3}$ times more?

_____

21. A tailor has $15\frac{3}{4}$ yards of fabric on a bolt. He wants to divide the fabric equally among three projects. How much fabric will he use for each project?

_____

22. Yuki worked $6\frac{1}{4}$ hours of overtime on Saturday. She spent $\frac{1}{3}$ of the time repairing a computer printer. What part of her overtime was spent repairing the printer?

_____

23. A recipe calls for $\frac{1}{3}$ of a cup of raisins. If you make four times the amount of the recipe, how many cups of raisins will you need?

_____

24. A survey found that $\frac{3}{10}$ of the people in an apartment complex own at least one cat. Of those, $\frac{1}{6}$ owned more than one cat. What fraction of the people in the apartment complex owned more than one cat?

_____

Check your answers on pages 85–86.

Add. Reduce to lowest terms.

1. $\dfrac{2}{7} + \dfrac{4}{7}$ _____

2. $\dfrac{3}{10} + \dfrac{1}{10}$ _____

3. $\dfrac{2}{5} + \dfrac{3}{10}$ _____

4. $\dfrac{3}{4} + \dfrac{1}{6}$ _____

5. $4\dfrac{3}{8} + 3\dfrac{3}{8}$ _____

6. $2\dfrac{3}{5} + 5\dfrac{4}{5}$ _____

7. $3\dfrac{1}{3} + 4\dfrac{1}{4}$ _____

8. $2\dfrac{5}{6} + 1\dfrac{2}{3}$ _____

Subtract these fractions. Reduce to lowest terms.

9. $\dfrac{8}{9} - \dfrac{3}{9}$ _____

10. $\dfrac{5}{6} - \dfrac{1}{6}$ _____

11. $\dfrac{3}{4} - \dfrac{1}{2}$ _____

12. $\dfrac{5}{6} - \dfrac{2}{3}$ _____

13. $4\dfrac{2}{3} - 2\dfrac{1}{3}$ _____

14. $8\dfrac{5}{6} - 5\dfrac{1}{6}$ _____

15. $6\dfrac{3}{4} - 2\dfrac{3}{8}$ _____

16. $2\dfrac{5}{6} - 1\dfrac{3}{4}$ _____

17. $4\dfrac{3}{5} - 1\dfrac{4}{5}$ _____

18. $8\dfrac{3}{8} - 4\dfrac{5}{8}$ _____

19. $5\dfrac{1}{6} - 3\dfrac{1}{2}$ _____

20. $6\dfrac{1}{4} - 1\dfrac{1}{3}$ _____

Multiply. Reduce to lowest terms.

21. $\dfrac{3}{4} \times \dfrac{2}{3}$ _____

22. $\dfrac{3}{10} \times \dfrac{2}{5}$ _____

23. $\dfrac{5}{8} \times \dfrac{4}{9}$ _____

24. $2\dfrac{2}{3} \times \dfrac{7}{8}$ _____

25. $5\dfrac{1}{2} \times 1\dfrac{1}{3}$ _____

26. $3\dfrac{3}{4} \times 3\dfrac{1}{5}$ _____

73

Divide. Reduce to lowest terms.

27. $\dfrac{1}{3} \div \dfrac{1}{6}$ _____

28. $\dfrac{5}{8} \div \dfrac{3}{10}$ _____

29. $\dfrac{3}{4} \div 6$ _____

30. $3\dfrac{1}{3} \div \dfrac{3}{7}$ _____

31. $5 \div 2\dfrac{1}{3}$ _____

32. $8\dfrac{1}{3} \div 1\dfrac{1}{4}$ _____

Solving Problems

Solve. Reduce your answers to lowest terms.

33. Carmen had two cans of tuna fish. One can held $6\dfrac{1}{3}$ ounces. The her can held $3\dfrac{1}{4}$ ounces. How much tuna fish did Carmen have all together?

_____

34. State Building Supplies had $45\dfrac{3}{8}$ pounds of concrete. They sold $12\dfrac{3}{4}$ pounds. How much concrete did they have left?

_____

35. Tom is able to plant $\dfrac{3}{4}$ of an acre a day. How many acres can Tom plant in $4\dfrac{1}{2}$ days?

_____

36. Keiko has a bottle containing $12\dfrac{1}{2}$ ounces of medicine. If she takes $\dfrac{5}{8}$ of an ounce at a time, how many times will she take the medicine?

_____

37. Sam is driving from Las Vegas to Phoenix. If he drives an average speed of 50 miles per hour, how far can he drive in $1\dfrac{3}{4}$ hours?

_____

38. A park contains $1\dfrac{2}{3}$ acres of land. According to the plans, $\dfrac{1}{4}$ of the park will be used for children's play equipment. How many acres will be set aside for children's play equipment?

_____

Check your answers on pages 86–87.

Answer Key

Lesson 1 ······▶ **WHAT ARE FRACTIONS?** (PAGES 2–3)

1. $\dfrac{2}{3}$ 2. 8; $\dfrac{5}{8}$ 3. $\dfrac{1}{2}$ 4. $\dfrac{3}{8}$ 5. $\dfrac{2}{3}$ 6. $\dfrac{4}{5}$ 7. $\dfrac{3}{4}$

8. $\dfrac{1}{3}$ 9. $\dfrac{2}{5}$ 10. $\dfrac{1}{2}$ 11. $\dfrac{5}{8}$ 12. $\dfrac{9}{10}$ 13. $\dfrac{3}{8}$ 14. $\dfrac{5}{8}$

Lesson 2 ······▶ **EQUIVALENT FRACTIONS** (PAGE 5)

1. 16 yes   2. $5 \times 5 = 25$ no

3. $3 \times 8 = 24$ $6 \times 4 = 24$ yes   4. $1 \times 8 = 8$ $3 \times 5 = 15$ no

5. $1 \times 9 = 9$ $3 \times 3 = 9$ yes   6. $4 \times 2 = 8$ $8 \times 1 = 8$ yes

7. $2 \times 9 = 18$ $5 \times 4 = 20$ no   8. $2 \times 5 = 10$ $10 \times 1 = 10$ yes

9. $2 \times 4 = 8$ $8 \times 1 = 8$ yes   10. $2 \times 6 = 12$ $4 \times 3 = 12$ yes

11. Yes $2 \times 4 = 8$ $8 \times 1 = 8$ 12. No $3 \times 10 = 30$ $5 \times 5 = 25$

Lesson 3 ······▶ **RAISING TO HIGHER TERMS** (PAGES 7–8)

1. $\dfrac{3}{9}$ 2. $4 \times 2 = 8$; $\dfrac{2}{8}$ 3. $\dfrac{2}{4}$ 4. $\dfrac{3}{6}$

5. $\dfrac{4}{8}$ 6. $\dfrac{5}{10}$ 7. $\dfrac{2}{6}$ 8. $\dfrac{4}{6}$

9. $\dfrac{6}{9}$ 10. $\dfrac{6}{8}$ 11. $\dfrac{4}{6}$; $\dfrac{6}{9}$ 12. $\dfrac{2}{8}$

Lesson 4 ······▶ **REDUCING TO LOWEST TERMS** (PAGE 10)

1. 3; $\dfrac{2}{3}$ 2. 3; 3; $\dfrac{2}{3}$ 3. $\dfrac{1}{4}$ 4. $\dfrac{1}{2}$

5. $\dfrac{7}{8}$ 6. $\dfrac{3}{4}$ 7. $\dfrac{2}{5}$ 8. $\dfrac{4}{5}$

9. $\dfrac{3}{5}$ 10. $\dfrac{1}{3}$ 11. Yes $\dfrac{1}{5}$ 12. No

Lesson 5 ·····▶ ADDING LIKE FRACTIONS (page 12)

1. $3; \dfrac{3}{5}$

2. $1 + 3 = 4 \quad \dfrac{4}{10} = \dfrac{2}{5}$

3. $\dfrac{2}{3}$

4. $\dfrac{3}{4}$

5. $\dfrac{4}{8} = \dfrac{1}{2}$

6. $\dfrac{4}{5}$

7. $\dfrac{5}{6}$

8. $\dfrac{6}{8} = \dfrac{3}{4}$

9. $\dfrac{6}{9} = \dfrac{2}{3}$

10. $\dfrac{3}{4}$

11. $\dfrac{4}{8} = \dfrac{1}{2}$ inch

12. $\dfrac{5}{6}$

Lesson 6 ·····▶ SUBTRACTING LIKE FRACTIONS (pages 13-14)

1. $\dfrac{3}{6} = \dfrac{1}{2}$

2. $6 - 3 = 3 \quad \dfrac{3}{9} = \dfrac{1}{3}$

3. $\dfrac{2}{8} = \dfrac{1}{4}$

4. $\dfrac{1}{4}$

5. $\dfrac{2}{5}$

6. $\dfrac{4}{8} = \dfrac{1}{2}$

7. $\dfrac{3}{6} = \dfrac{1}{2}$

8. $\dfrac{3}{9} = \dfrac{1}{3}$

9. $\dfrac{4}{10} = \dfrac{2}{5}$

10. $\dfrac{1}{3}$

11. $\dfrac{2}{8} = \dfrac{1}{4}$ yard

12. $\dfrac{6}{10} = \dfrac{3}{5}$ ounce

Lesson 7 ·····▶ FINDING A COMMON DENOMINATOR (pages 16-17)

1. 2
2. 2, 12
3. 12
4. 8
5. 6
6. 4
7. 10
8. 15
9. 20
10. 12
11. 8
12. 20

Lesson 8 ·····▶ ADDING UNLIKE FRACTIONS (pages 19-20)

1. $\dfrac{3}{4}$

2. $12 \quad 12 \quad \dfrac{8}{12} + \dfrac{3}{12} = \dfrac{11}{12}$

3. $\dfrac{3}{12} + \dfrac{4}{12} = \dfrac{7}{12}$

4. $\dfrac{5}{10} + \dfrac{4}{10} = \dfrac{9}{10}$

5. $\dfrac{5}{20} + \dfrac{4}{20} = \dfrac{9}{20}$

6. $\dfrac{1}{6} + \dfrac{2}{6} = \dfrac{3}{6}$ reduces to $\dfrac{1}{2}$

7. $\dfrac{2}{8} + \dfrac{3}{8} = \dfrac{5}{8}$

8. $\dfrac{3}{24} + \dfrac{16}{24} = \dfrac{19}{24}$

9. $\dfrac{3}{8} + \dfrac{4}{8} = \dfrac{7}{8}$

10. $\dfrac{12}{30} + \dfrac{5}{30} = \dfrac{17}{30}$

11. $\dfrac{6}{8} + \dfrac{1}{8} = \dfrac{7}{8}$ pound

12. $\dfrac{3}{15} + \dfrac{10}{15} = \dfrac{13}{15}$ mile

Lesson 9 ·····▶ SUBTRACTING UNLIKE FRACTIONS (pages 21-22)

1. $\dfrac{5}{8}$

2. $\dfrac{1 \times 4}{3 \times 4} = \dfrac{4}{12} \quad \dfrac{1 \times 3}{4 \times 3} = \dfrac{3}{12} \quad \dfrac{4}{12} - \dfrac{3}{12} = \dfrac{1}{12}$

3. $\dfrac{2}{4} - \dfrac{1}{4} = \dfrac{1}{4}$

4. $\dfrac{4}{6} - \dfrac{1}{6} = \dfrac{3}{6}$ reduces to $\dfrac{1}{2}$

5. $\dfrac{5}{8} - \dfrac{2}{8} = \dfrac{3}{8}$ 6. $\dfrac{7}{8} - \dfrac{6}{8} = \dfrac{1}{8}$

7. $\dfrac{8}{12} - \dfrac{3}{12} = \dfrac{5}{12}$ 8. $\dfrac{9}{15} - \dfrac{5}{15} = \dfrac{4}{15}$

9. $\dfrac{15}{20} - \dfrac{8}{20} = \dfrac{7}{20}$ 10. $\dfrac{16}{24} - \dfrac{15}{24} = \dfrac{1}{24}$

11. $\dfrac{6}{8} - \dfrac{1}{8} = \dfrac{5}{8}$ cup   12. $\dfrac{15}{24} - \dfrac{8}{24} = \dfrac{7}{24}$ pound

Lesson 10 ·····► CHANGING MIXED NUMBERS TO IMPROPER FRACTIONS (PAGES 24–25)

1. $\dfrac{27}{8}$

2. $12 + 2 = 14$ $4\dfrac{2}{3} = \dfrac{14}{3}$

3. $\dfrac{9}{4}$

4. $\dfrac{13}{8}$

5. $\dfrac{11}{3}$

6. $\dfrac{13}{5}$

7. $\dfrac{49}{6}$

8. $\dfrac{32}{5}$

9. $\dfrac{7}{4}$

10. $\dfrac{47}{8}$

11. $\dfrac{11}{3}$

12. $\dfrac{8}{3}$

Lesson 11 ·····► CHANGING IMPROPER FRACTIONS TO MIXED NUMBERS (PAGES 27–28)

1. $4\dfrac{2}{3}$

2. $4, 4$

3. $2\dfrac{1}{2}$

4. $2\dfrac{1}{3}$

5. $2\dfrac{2}{4} = 2\dfrac{1}{2}$

6. 3

7. $3\dfrac{1}{2}$

8. $2\dfrac{2}{3}$

9. $2\dfrac{2}{5}$

10. $5\dfrac{3}{4}$

11. $2\dfrac{1}{8}$

12. $5\dfrac{1}{4}$

Lesson 12 ·····► ADDING MIXED NUMBERS WITH LIKE FRACTIONS (PAGE 30)

1. $\dfrac{3}{5}$; $5\dfrac{3}{5}$

2. $1 + 4 = 5$ $\dfrac{12}{8} = 1\dfrac{4}{8}$ $\dfrac{4}{8} = \dfrac{1}{2}$ The answer is $6\dfrac{1}{2}$.

3. $5\dfrac{2}{3}$

4. $3\dfrac{2}{4}$ reduces to $3\dfrac{1}{2}$

5. $5\frac{3}{6}$ reduces to $5\frac{1}{2}$ 6. $5\frac{3}{4}$

7. $2\frac{4}{5}$ 8. $7\frac{6}{8}$ reduces to $7\frac{3}{4}$

9. $4\frac{6}{5} = 4 + 1 + \frac{1}{5} = 5\frac{1}{5}$ 10. $8\frac{10}{6} = 8 + 1 + \frac{4}{6} = 9\frac{4}{6}$ reduces to $9\frac{2}{3}$

11. $3\frac{4}{8}$ reduces to $3\frac{1}{2}$ 12. $8\frac{10}{8} = 8 + 1 + \frac{2}{8} = 9\frac{2}{8}$ reduces to $9\frac{1}{4}$

Lesson 13·······▶ ADDING MIXED NUMBERS WITH UNLIKE FRACTIONS
(PAGES 32–33)

1. $\frac{2}{6} + \frac{3}{6} = \frac{5}{6}$ The answer is $3\frac{5}{6}$.

2. $2 + 1 = 3$ $\frac{5}{4} = 1\frac{1}{4}$ $3 + 1\frac{1}{4} = 4\frac{1}{4}$ The answer is $4\frac{1}{4}$.

3. $2\frac{2}{6} + 1\frac{3}{6} = 3\frac{5}{6}$

4. $1\frac{1}{8} + 1\frac{2}{8} = 2\frac{3}{8}$

5. $3\frac{2}{4} + 1\frac{3}{4} = 4\frac{5}{4} = 4 + 1 + \frac{1}{4} = 5\frac{1}{4}$

6. $1\frac{2}{6} + 2\frac{5}{6} = 3\frac{7}{6} = 3 + 1 + \frac{1}{6} = 4\frac{1}{6}$

7. $2\frac{3}{8} + 1\frac{4}{8} = 3\frac{7}{8}$

8. $1\frac{4}{12} + 2\frac{3}{12} = 3\frac{7}{12}$

9. $2\frac{6}{8} + 3\frac{7}{8} = 5\frac{13}{8} = 5 + 1 + \frac{5}{8} = 6\frac{5}{8}$

10. $1\frac{3}{12} + 2\frac{10}{12} = 3\frac{13}{12} = 3 + 1 + \frac{1}{12} = 4\frac{1}{12}$

11. $3\frac{2}{4} + 2\frac{3}{4} = 5\frac{5}{4} = 5 + 1 + \frac{1}{4} = 6\frac{1}{4}$

12. $1\frac{5}{10} + 2\frac{2}{10} = 3\frac{7}{10}$

Lesson 14 ·····▶ SUBTRACTING MIXED NUMBERS WITH LIKE FRACTIONS (PAGES 34–35)

1. $\frac{2}{3}$ The answer is $1\frac{2}{3}$.

2. $5 - 3 = 2$ $\frac{4}{5} - \frac{3}{5} = \frac{1}{5}$ The answer is $2\frac{1}{5}$.

3. $1\frac{2}{4}$ reduces to $1\frac{1}{2}$

4. $2\frac{4}{6}$ reduces to $2\frac{2}{3}$

5. $2\frac{4}{8}$ reduces to $2\frac{1}{2}$

6. $1\frac{1}{5}$

7. $3\frac{3}{9}$ reduces to $3\frac{1}{3}$

8. $1\frac{1}{3}$

9. $4\frac{2}{8}$ reduces to $4\frac{1}{4}$

10. $3\frac{2}{4},$ reduces to $3\frac{1}{2}$

11. $6\frac{2}{5}$

12. $2\frac{2}{4}$ reduces to $2\frac{1}{2}$

Lesson 15 ·····▶ SUBTRACTING MIXED NUMBERS WITH UNLIKE FRACTIONS (PAGES 37–38)

1. $\frac{3}{6} = \frac{1}{2}$ The answer is $2\frac{1}{2}$.

2. $5 - 2 = 3$ $\frac{8}{12} - \frac{3}{12} = \frac{5}{12}$ The answer is $3\frac{5}{12}$.

3. $1\frac{2}{4} - 1\frac{1}{4} = \frac{1}{4}$

4. $2\frac{5}{6} - 1\frac{4}{6} = 1\frac{1}{6}$

5. $3\frac{5}{8} - 1\frac{2}{8} = 2\frac{3}{8}$

6. $5\frac{6}{8} - 2\frac{1}{8} = 3\frac{5}{8}$

7. $4\frac{4}{6} - 1\frac{3}{6} = 3\frac{1}{6}$

8. $3\frac{4}{12} - 2\frac{3}{12} = 1\frac{1}{12}$

9. $6\frac{6}{15} - 3\frac{5}{15} = 3\frac{1}{15}$

10. $4\frac{10}{12} - 3\frac{9}{12} = 1\frac{1}{12}$

11. $3\frac{2}{4} - 1\frac{1}{4} = 2\frac{1}{4}$

12. $3\frac{4}{6} - 1\frac{3}{6} = 2\frac{1}{6}$

Lesson 16 ·····▶ SUBTRACTING LIKE MIXED NUMBERS WITH BORROWING (PAGES 40–41)

1. $\frac{4}{3} - \frac{2}{3} = \frac{2}{3}$ The answer is $1\frac{2}{3}$.

2. $4 - 2 = 2$ $\frac{8}{8} - \frac{3}{8} = \frac{5}{8}$ The answer is $2\frac{5}{8}$.

3. $1\frac{8}{5} - 1\frac{4}{5} = \frac{4}{5}$

4. $2\frac{4}{4} - \frac{3}{4} = 2\frac{1}{4}$

5. $4\frac{9}{8} - 3\frac{2}{8} = 1\frac{7}{8}$

6. $3\frac{7}{6} - 2\frac{5}{6} = 1\frac{2}{6}$ reduces to $1\frac{1}{3}$

7. $7\frac{5}{5} - 3\frac{3}{5} = 4\frac{2}{5}$

8. $4\frac{6}{6} - 1\frac{5}{6} = 3\frac{1}{6}$

9. $2\frac{4}{3} - 1\frac{2}{3} = 1\frac{2}{3}$

10. $8\frac{2}{2} - 5\frac{1}{2} = 3\frac{1}{2}$

11. $4\frac{13}{8} - 2\frac{7}{8} = 2\frac{6}{8}$ reduces to $2\frac{3}{4}$ feet

12. $7\frac{4}{4} - \frac{3}{4} = 7\frac{1}{4}$ pounds

Lesson 17 ·······► SUBTRACTING UNLIKE MIXED NUMBERS WITH BORROWING (PAGES 43–44)

1. 3; $1\frac{3}{4}$

2. $\frac{15}{12} - \frac{4}{12} = \frac{11}{12}$ The answer is $1\frac{11}{12}$

3. $2\frac{5}{4} - 1\frac{2}{4} = 1\frac{3}{4}$

4. $3\frac{11}{8} - 1\frac{6}{8} = 2\frac{5}{8}$

5. $2\frac{10}{6} - 1\frac{5}{6} = 1\frac{5}{6}$

6. $3\frac{10}{8} - 2\frac{3}{8} = 1\frac{7}{8}$

7. $4\frac{16}{12} - 2\frac{9}{12} = 2\frac{7}{12}$

8. $4\frac{14}{12} - 1\frac{9}{12} = 3\frac{5}{12}$

9. $5\frac{13}{10} - 2\frac{6}{10} = 3\frac{7}{10}$

10. $7\frac{28}{20} - 5\frac{15}{20} = 2\frac{13}{20}$

11. $4\frac{6}{4} - 1\frac{3}{4} = 3\frac{3}{4}$

12. $11\frac{18}{15} - 8\frac{5}{15} = 3\frac{13}{15}$

Lesson 18 ·······► REVIEW OF ADDITION AND SUBTRACTION (PAGES 45–46)

1. $\frac{4}{8}$ reduces to $\frac{1}{2}$

2. $\frac{3}{5}$

3. $\frac{6}{9}$ reduces to $\frac{2}{3}$

4. $\frac{4}{6} + \frac{1}{6} = \frac{5}{6}$

5. $\frac{3}{8} + \frac{2}{8} = \frac{5}{8}$

6. $\frac{2}{9} + \frac{6}{9} = \frac{8}{9}$

7. $3\frac{3}{3} = 4$

8. $5\frac{3}{5}$

9. $3\frac{1}{4} + 1\frac{2}{4} = 4\frac{3}{4}$

10. $2\frac{9}{12} + 2\frac{8}{12} = 4\frac{17}{12} = 5\frac{5}{12}$

11. $\dfrac{4}{6}$ reduces to $\dfrac{2}{3}$

12. $\dfrac{2}{4}$ reduces to $\dfrac{1}{2}$

13. $\dfrac{4}{8}$ reduces to $\dfrac{1}{2}$

14. $\dfrac{5}{8} - \dfrac{2}{8} = \dfrac{3}{8}$

15. $\dfrac{5}{6} - \dfrac{4}{6} = \dfrac{1}{6}$

16. $2\dfrac{2}{4}$ reduces to $2\dfrac{1}{2}$

17. $3\dfrac{4}{8}$ reduces to $3\dfrac{1}{2}$

18. $4\dfrac{4}{6} - 1\dfrac{1}{6} = 3\dfrac{3}{6}$ reduces to $3\dfrac{1}{2}$

19. $3\dfrac{9}{12} - 2\dfrac{8}{12} = 1\dfrac{1}{12}$

20. $5\dfrac{5}{4} - 4\dfrac{3}{4} = 1\dfrac{2}{4}$ reduces to $1\dfrac{1}{2}$

21. $7\dfrac{7}{6} - 4\dfrac{5}{6} = 3\dfrac{2}{6}$ reduces to $3\dfrac{1}{3}$

22. $3\dfrac{5}{4} - 1\dfrac{2}{4} = 2\dfrac{3}{4}$

23. $5\dfrac{14}{12} - 3\dfrac{3}{12} = 2\dfrac{11}{12}$

24. $2\dfrac{8}{12} + 3\dfrac{9}{12} = 5\dfrac{17}{12} = 5 + 1 + \dfrac{5}{12} = 6\dfrac{5}{12}$ miles

25. $5\dfrac{1}{2} - \dfrac{3}{4} = 5\dfrac{2}{4} - \dfrac{3}{4} = 4\dfrac{6}{4} - \dfrac{3}{4} = 4\dfrac{3}{4}$ pounds

26. $1\dfrac{1}{2} + 2\dfrac{3}{4} = 1\dfrac{2}{4} + 2\dfrac{3}{4} = 3\dfrac{5}{4} = 4\dfrac{1}{4}$ hours

27. $12 - 8\dfrac{1}{2} = 11\dfrac{2}{2} - 8\dfrac{1}{2} = 3\dfrac{1}{2}$ ounces

28. $9\dfrac{1}{4} - 5\dfrac{7}{8} = 9\dfrac{2}{8} - 5\dfrac{7}{8} = 8\dfrac{10}{8} - 5\dfrac{7}{8} = 3\dfrac{3}{8}$ yards

29. $1\dfrac{1}{4} + \dfrac{1}{3} = 1\dfrac{3}{12} + \dfrac{4}{12} = 1\dfrac{7}{12}$ hour

Lesson 19······▶ FINDING FRACTIONAL PARTS (PAGES 47–48)

1. $33 \div 4 = 8\dfrac{1}{4}$

2. $4\dfrac{4}{8} = 4\dfrac{1}{2}$

3. $\dfrac{8}{4} = 2$

4. $\dfrac{7}{3} = 2\dfrac{1}{3}$

5. $\dfrac{18}{3} = 6$

6. $\dfrac{22}{3} = 7\dfrac{1}{3}$

7. $\dfrac{15}{6} = 2\dfrac{3}{6}$ reduces to $2\dfrac{1}{2}$

8. $\dfrac{36}{5} = 7\dfrac{1}{5}$

9. $\dfrac{21}{8} = 2\dfrac{5}{8}$

10. $\dfrac{50}{6} = 8\dfrac{2}{6}$ reduces to $8\dfrac{1}{3}$

11. $\dfrac{8}{4} = 2$ hours

12. $\dfrac{50}{5} = 10$ doubles

Lesson 20 ·······▶ MULTIPLYING PROPER FRACTIONS (PAGES 49–50)

1. $\dfrac{3}{8}$

2. $2 \times 1 = 2$ $3 \times 2 = 6$ $\dfrac{2}{6} = \dfrac{1}{3}$

3. $\dfrac{1}{8}$

4. $\dfrac{3}{12}$ reduces to $\dfrac{1}{4}$

5. $\dfrac{6}{12}$ reduces to $\dfrac{1}{2}$

6. $\dfrac{1}{20}$

7. $\dfrac{6}{15}$ reduces to $\dfrac{2}{5}$

8. $\dfrac{6}{20}$ reduces to $\dfrac{3}{10}$

9. $\dfrac{2}{18}$ reduces to $\dfrac{1}{9}$

10. $\dfrac{15}{24}$ reduces to $\dfrac{5}{8}$

11. $\dfrac{3}{8}$ of a loaf

12. $\dfrac{8}{15}$ mile

Lesson 21 ·······▶ CANCELING (PAGES 52–53)

1. $\dfrac{1}{2} \times \dfrac{1}{1} = \dfrac{1}{2}$

2. $\dfrac{3}{8} \times \dfrac{5}{9} = \dfrac{5}{24}$

3. $\dfrac{1}{2} \times \dfrac{1}{1} = \dfrac{1}{2}$

4. $\dfrac{1}{4} \times \dfrac{1}{2} = \dfrac{1}{8}$

5. $\dfrac{1}{1} \times \dfrac{1}{5} = \dfrac{1}{5}$

6. $\dfrac{1}{4} \times \dfrac{1}{1} = \dfrac{1}{4}$

7. $\dfrac{1}{3} \times \dfrac{1}{2} = \dfrac{1}{6}$

8. $\dfrac{1}{2} \times \dfrac{1}{1} = \dfrac{1}{2}$

9. $\dfrac{1}{2} \times \dfrac{1}{1} = \dfrac{1}{2}$

10. $\dfrac{1}{3} \times \dfrac{1}{1} = \dfrac{1}{3}$

11. $\dfrac{1}{1} \times \dfrac{1}{8} = \dfrac{1}{8}$

12. $\dfrac{5}{3} \times \dfrac{1}{3} = \dfrac{5}{9}$ of a pound

Lesson 22 ·······▶ MULTIPLYING MIXED NUMBERS (PAGES 55–56)

1. $3\dfrac{3}{4}$

2. $\dfrac{7}{2} = 3\dfrac{1}{2}$

3. $\dfrac{8}{3} \times \dfrac{5}{4} = \dfrac{2}{3} \times \dfrac{5}{1} = \dfrac{10}{3} = 3\dfrac{1}{3}$

4. $\dfrac{3}{2} \times \dfrac{3}{2} = \dfrac{9}{4} = 2\dfrac{1}{4}$

5. $\dfrac{5}{4} \times \dfrac{7}{3} = \dfrac{35}{12} = 2\dfrac{11}{12}$

6. $\dfrac{9}{8} \times \dfrac{5}{3} = \dfrac{3}{8} \times \dfrac{5}{1} = \dfrac{15}{8} = 1\dfrac{7}{8}$

7. $\dfrac{7}{6} \times \dfrac{3}{2} = \dfrac{7}{2} \times \dfrac{1}{2} = \dfrac{7}{4} = 1\dfrac{3}{4}$

8. $\dfrac{9}{4} \times \dfrac{10}{7} = \dfrac{9}{2} \times \dfrac{5}{7} = \dfrac{45}{14} = 3\dfrac{3}{14}$

9. $\dfrac{11}{6} \times \dfrac{9}{4} = \dfrac{11}{2} \times \dfrac{3}{4} = \dfrac{33}{8} = 4\dfrac{1}{8}$

10. $\dfrac{20}{9} \times \dfrac{15}{8} = \dfrac{5}{3} \times \dfrac{5}{2} = \dfrac{25}{6} = 4\dfrac{1}{6}$

11. $\dfrac{7}{2} \times \dfrac{15}{7} = \dfrac{1}{2} \times \dfrac{15}{1} = \dfrac{15}{2} = 7\dfrac{1}{2}$ feet

12. $\dfrac{5}{2} \times \dfrac{18}{5} = \dfrac{1}{1} \times \dfrac{9}{1} = 9$ gallons

Lesson 23 ·····▶ THE RELATIONSHIP BETWEEN MULTIPLICATION AND DIVISION (PAGE 58)

1. $\frac{4}{3}$ is the reciprocal of $\frac{3}{4}$.

2. $\frac{6}{1}$; $\frac{1}{6}$ is the reciprocal of $\frac{6}{1}$

3. $\frac{2}{1}$

4. $\frac{8}{3}$

5. $\frac{10}{9}$

6. $\frac{3}{2}$

7. $\frac{5}{8}$

8. $\frac{4}{9}$

9. $\frac{1}{4}$

10. $\frac{1}{9}$

11. $\frac{3}{2}$

12. $\frac{4}{3}$

Lesson 24 ·····▶ DIVIDING BY PROPER FRACTIONS (PAGES 60–61)

1. $\frac{25}{2} = 12\frac{1}{2}$

2. $\frac{15}{8} = 1\frac{7}{8}$

3. $\frac{1}{2} \times \frac{4}{1} = \frac{1}{1} \times \frac{2}{1} = \frac{2}{1} = 2$

4. $\frac{1}{3} \times \frac{6}{5} = \frac{1}{1} \times \frac{2}{5} = \frac{2}{5}$

5. $\frac{2}{3} \times \frac{4}{3} = \frac{8}{9}$

6. $\frac{3}{4} \times \frac{3}{2} = \frac{9}{8} = 1\frac{1}{8}$

7. $\frac{18}{1} \times \frac{4}{3} = \frac{6}{1} \times \frac{4}{1} = \frac{24}{1} = 24$

8. $\frac{15}{1} \times \frac{8}{5} = \frac{3}{1} \times \frac{8}{1} = \frac{24}{1} = 24$

9. $\frac{8}{1} \times \frac{5}{1} = \frac{40}{1} = 40$

10. $\frac{10}{1} \times \frac{3}{2} = \frac{5}{1} \times \frac{3}{1} = \frac{15}{1} = 15$

11. $8 \div \frac{2}{3} = \frac{8}{1} \times \frac{3}{2} = \frac{4}{1} \times \frac{3}{1} = 12$

12. $\frac{5}{6} \div \frac{3}{8} = \frac{5}{6} \times \frac{8}{3} = \frac{5}{3} \times \frac{4}{3} = \frac{20}{9} = 2\frac{2}{9}$ bowls

Lesson 25 ·····▶ DIVIDING BY A WHOLE NUMBER (PAGES 63–64)

1. $\frac{1}{8}$

2. The reciprocal of $\frac{3}{1}$ is $\frac{1}{3}$. $\frac{5}{6} \times \frac{1}{3} = \frac{5}{18}$

3. $\frac{1}{2} \times \frac{1}{4} = \frac{1}{8}$

4. $\frac{2}{3} \times \frac{1}{4} = \frac{1}{3} \times \frac{1}{2} = \frac{1}{6}$

5. $\frac{1}{6} \times \frac{1}{2} = \frac{1}{12}$

6. $\frac{3}{4} \times \frac{1}{6} = \frac{1}{4} \times \frac{1}{2} = \frac{1}{8}$

7. $\dfrac{5}{6} \times \dfrac{1}{5} = \dfrac{1}{6} \times \dfrac{1}{1} = \dfrac{1}{6}$

8. $\dfrac{2}{3} \times \dfrac{1}{8} = \dfrac{1}{3} \times \dfrac{1}{4} = \dfrac{1}{12}$

9. $\dfrac{4}{5} \times \dfrac{1}{3} = \dfrac{4}{15}$

10. $\dfrac{5}{8} \times \dfrac{1}{10} = \dfrac{1}{8} \times \dfrac{1}{2} = \dfrac{1}{16}$

11. $\dfrac{3}{4} \div 6 = \dfrac{3}{4} \times \dfrac{1}{6} = \dfrac{1}{4} \times \dfrac{1}{2} = \dfrac{1}{8}$ cup

12. $\dfrac{3}{4} \div 15 = \dfrac{3}{4} \times \dfrac{1}{15} = \dfrac{1}{4} \times \dfrac{1}{5} = \dfrac{1}{20}$ pound

Lesson 26 ·······▶ DIVIDING MIXED NUMBERS (PAGES 66–67)

1. $\dfrac{4}{9}$

2. $\dfrac{45}{8}$; $\dfrac{45}{8} = 5\dfrac{5}{8}$

3. $\dfrac{5}{4} \div \dfrac{1}{2} = \dfrac{5}{4} \times \dfrac{2}{1} = \dfrac{5}{2} \times \dfrac{1}{1} = \dfrac{5}{2} = 2\dfrac{1}{2}$

4. $\dfrac{8}{3} \div 4 = \dfrac{8}{3} \times \dfrac{1}{4} = \dfrac{2}{3} \times \dfrac{1}{1} = \dfrac{2}{3}$

5. $\dfrac{4}{3} \div \dfrac{3}{4} = \dfrac{4}{3} \times \dfrac{4}{3} = \dfrac{16}{9} = 1\dfrac{7}{9}$

6. $\dfrac{12}{5} \div 6 = \dfrac{12}{5} \times \dfrac{1}{6} = \dfrac{2}{5} \times \dfrac{1}{1} = \dfrac{2}{5}$

7. $\dfrac{15}{4} \div \dfrac{1}{5} = \dfrac{15}{4} \times \dfrac{5}{1} = \dfrac{75}{4} = 18\dfrac{3}{4}$

8. $\dfrac{11}{2} \div \dfrac{4}{5} = \dfrac{11}{2} \times \dfrac{5}{4} = \dfrac{55}{8} = 6\dfrac{7}{8}$

9. $\dfrac{11}{6} \div \dfrac{1}{2} = \dfrac{11}{6} \times \dfrac{2}{1} = \dfrac{11}{3} \times \dfrac{1}{1} = \dfrac{11}{3} = 3\dfrac{2}{3}$

10. $\dfrac{14}{5} \div \dfrac{3}{10} = \dfrac{14}{5} \times \dfrac{10}{3} = \dfrac{14}{1} \times \dfrac{2}{3} = \dfrac{28}{3} = 9\dfrac{1}{3}$

11. $4\dfrac{1}{2} \div \dfrac{3}{4} = \dfrac{9}{2} \times \dfrac{4}{3} = \dfrac{3}{1} \times \dfrac{2}{1} = \dfrac{6}{1} = 6$

12. $10\dfrac{1}{8} \div 9 = \dfrac{81}{8} \times \dfrac{1}{9} = \dfrac{9}{8} \times \dfrac{1}{1} = \dfrac{9}{8} = 1\dfrac{1}{8}$ ounce

Lesson 27 ·······▶ DIVIDING BY A MIXED NUMBER (PAGES 69–70)

1. $\dfrac{1}{2}$

2. $\dfrac{1}{3} \times \dfrac{5}{1} = \dfrac{5}{3}$; $\dfrac{5}{3} = 1\dfrac{2}{3}$

3. $3 \div \dfrac{3}{2} = \dfrac{3}{1} \times \dfrac{2}{3} = \dfrac{1}{1} \times \dfrac{2}{1} = \dfrac{2}{1} = 2$

4. $6 \div \dfrac{4}{3} = \dfrac{6}{1} \times \dfrac{3}{4} = \dfrac{3}{1} \times \dfrac{3}{2} = \dfrac{9}{2} = 4\dfrac{1}{2}$

5. $10 \div \dfrac{14}{5} = \dfrac{10}{1} \times \dfrac{5}{14} = \dfrac{5}{1} \times \dfrac{5}{7} = \dfrac{25}{7} = 3\dfrac{4}{7}$

6. $\dfrac{3}{4} \div \dfrac{9}{4} = \dfrac{3}{4} \times \dfrac{4}{9} = \dfrac{1}{1} \times \dfrac{1}{3} = \dfrac{1}{3}$

7. $\dfrac{5}{6} \div \dfrac{5}{3} = \dfrac{5}{6} \times \dfrac{3}{5} = \dfrac{1}{2} \times \dfrac{1}{1} = \dfrac{1}{2}$

8. $\dfrac{20}{3} \div \dfrac{5}{2} = \dfrac{20}{3} \times \dfrac{2}{5} = \dfrac{4}{3} \times \dfrac{2}{1} = \dfrac{8}{3} = 2\dfrac{2}{3}$

9. $\dfrac{22}{3} \div \dfrac{11}{4} = \dfrac{22}{3} \times \dfrac{4}{11} = \dfrac{2}{3} \times \dfrac{4}{1} = \dfrac{8}{3} = 2\dfrac{2}{3}$

10. $\dfrac{24}{5} \div \dfrac{5}{2} = \dfrac{24}{5} \times \dfrac{2}{5} = \dfrac{48}{25} = 1\dfrac{23}{25}$

11. $12 \div 1\dfrac{1}{2} = \dfrac{12}{1} \div \dfrac{3}{2} = \dfrac{12}{1} \times \dfrac{2}{3} = \dfrac{4}{1} \times \dfrac{2}{1} = 8$ horses

12. $12\dfrac{1}{2} \div 1\dfrac{1}{4} = \dfrac{25}{2} \div \dfrac{5}{4} = \dfrac{25}{2} \times \dfrac{4}{5} = \dfrac{5}{1} \times \dfrac{2}{1} = \dfrac{10}{1} = 10$ children

Lesson 28 ·····▶ REVIEW OF MULTIPLICATION AND DIVISION (PAGES 71-72)

1. $\dfrac{2}{5} \times \dfrac{1}{1} = \dfrac{2}{5}$

2. $\dfrac{1}{2} \times \dfrac{1}{1} = \dfrac{1}{2}$

3. $\dfrac{5}{1} \times \dfrac{1}{8} = \dfrac{5}{8}$

4. $\dfrac{1}{1} \times \dfrac{2}{9} = \dfrac{2}{9}$

5. $\dfrac{5}{4} \times \dfrac{7}{3} = \dfrac{35}{12} = 2\dfrac{11}{12}$

6. $\dfrac{18}{5} \times \dfrac{20}{9} = \dfrac{2}{1} \times \dfrac{4}{1} = \dfrac{8}{1} = 8$

7. $\dfrac{5}{2} \times \dfrac{14}{5} = \dfrac{1}{1} \times \dfrac{7}{1} = \dfrac{7}{1} = 7$

8. $\dfrac{35}{8} \times \dfrac{18}{7} = \dfrac{5}{4} \times \dfrac{9}{1} = \dfrac{45}{4} = 11\dfrac{1}{4}$

9. $\dfrac{3}{4} \times \dfrac{5}{3} = \dfrac{1}{4} \times \dfrac{5}{1} = \dfrac{5}{4} = 1\dfrac{1}{4}$

10. $\dfrac{5}{8} \times \dfrac{4}{3} = \dfrac{5}{2} \times \dfrac{1}{3} = \dfrac{5}{6}$

11. $\dfrac{2}{3} \times \dfrac{7}{6} = \dfrac{1}{3} \times \dfrac{7}{3} = \dfrac{7}{9}$

12. $\dfrac{8}{1} \times \dfrac{4}{3} = \dfrac{32}{3} = 10\dfrac{2}{3}$

13. $\dfrac{4}{1} \times \dfrac{3}{2} = \dfrac{2}{1} \times \dfrac{3}{1} = \dfrac{6}{1} = 6$

14. $\dfrac{5}{6} \times \dfrac{1}{2} = \dfrac{5}{12}$

15. $\dfrac{4}{5} \times \dfrac{1}{6} = \dfrac{2}{5} \times \dfrac{1}{3} = \dfrac{2}{15}$

16. $\dfrac{8}{3} \times \dfrac{4}{1} = \dfrac{32}{3} = 10\dfrac{2}{3}$

17. $\dfrac{16}{1} \times \dfrac{3}{8} = \dfrac{2}{1} \times \dfrac{3}{1} = \dfrac{6}{1} = 6$

18. $\dfrac{5}{6} \times \dfrac{7}{10} = \dfrac{1}{6} \times \dfrac{7}{2} = \dfrac{7}{12}$

19. $10\dfrac{1}{4} \div 1\dfrac{1}{2} = \dfrac{41}{4} \div \dfrac{3}{2} = \dfrac{41}{4} \times \dfrac{2}{3} = \dfrac{41}{2} \times \dfrac{1}{3} = \dfrac{41}{6} = 6\dfrac{5}{6}$ miles per gallon

20. $4\dfrac{1}{2} \times 2\dfrac{1}{3} = \dfrac{9}{2} \times \dfrac{7}{3} = \dfrac{3}{2} \times \dfrac{7}{1} = \dfrac{21}{2} = 10\dfrac{1}{2}$ tablespoons

21. $15\dfrac{3}{4} \div 3 = \dfrac{63}{4} \times \dfrac{1}{3} = \dfrac{21}{4} \times \dfrac{1}{1} = \dfrac{21}{4} = 5\dfrac{1}{4}$ yards

22. $6\dfrac{1}{4} \times \dfrac{1}{3} = \dfrac{25}{4} \times \dfrac{1}{3} = \dfrac{25}{12} = 2\dfrac{1}{12}$ hours

23. $\dfrac{1}{3} \times 4 = \dfrac{1}{3} \times \dfrac{4}{1} = \dfrac{4}{3} = 1\dfrac{1}{3}$ cups

24. $\dfrac{3}{10} \times \dfrac{1}{6} = \dfrac{1}{10} \times \dfrac{1}{2} = \dfrac{1}{20}$ of the people

Lesson 29 ······▶ MIXED REVIEW (PAGES 73–74)

1. $\dfrac{6}{7}$

2. $\dfrac{4}{10}$ reduces to $\dfrac{2}{5}$

3. $\dfrac{4}{10} + \dfrac{3}{10} = \dfrac{7}{10}$

4. $\dfrac{9}{12} + \dfrac{2}{12} = \dfrac{11}{12}$

5. $7\dfrac{6}{8}$ reduces to $7\dfrac{3}{4}$

6. $7\dfrac{7}{5} = 7 + 1 + \dfrac{2}{5} = 8\dfrac{2}{5}$

7. $3\dfrac{4}{12} + 4\dfrac{3}{12} = 7\dfrac{7}{12}$

8. $2\dfrac{5}{6} + 1\dfrac{4}{6} = 3\dfrac{9}{6} = 3 + 1 + \dfrac{3}{6} = 4\dfrac{3}{6}$ reduces to $4\dfrac{1}{2}$

9. $\dfrac{5}{9}$

10. $\dfrac{4}{6}$ reduces to $\dfrac{2}{3}$

11. $\dfrac{3}{4} - \dfrac{2}{4} = \dfrac{1}{4}$

12. $\dfrac{5}{6} - \dfrac{4}{6} = \dfrac{1}{6}$

13. $2\dfrac{1}{3}$

14. $3\dfrac{4}{6}$ reduces to $3\dfrac{2}{3}$

15. $6\dfrac{6}{8} - 2\dfrac{3}{8} = 4\dfrac{3}{8}$

16. $2\dfrac{10}{12} - 1\dfrac{9}{12} = 1\dfrac{1}{12}$

17. $3\dfrac{8}{5} - 1\dfrac{4}{5} = 2\dfrac{4}{5}$

18. $7\dfrac{11}{8} - 4\dfrac{5}{8} = 3\dfrac{6}{8}$ reduces to $3\dfrac{3}{4}$

19. $5\dfrac{1}{6} - 3\dfrac{3}{6} = 4\dfrac{7}{6} - 3\dfrac{3}{6} = 1\dfrac{4}{6}$ reduces to $1\dfrac{2}{3}$

20. $6\dfrac{3}{12} - 1\dfrac{4}{12} = 5\dfrac{15}{12} - 1\dfrac{4}{12} = 4\dfrac{11}{12}$

21. $\dfrac{1}{2} \times \dfrac{1}{1} = \dfrac{1}{2}$

22. $\dfrac{3}{5} \times \dfrac{1}{5} = \dfrac{3}{25}$

23. $\dfrac{5}{2} \times \dfrac{1}{9} = \dfrac{5}{18}$

24. $\dfrac{8}{3} \times \dfrac{7}{8} = \dfrac{1}{3} \times \dfrac{7}{1} = \dfrac{7}{3} = 2\dfrac{1}{3}$

25. $\dfrac{11}{2} \times \dfrac{4}{3} = \dfrac{11}{1} \times \dfrac{2}{3} = \dfrac{22}{3} = 7\dfrac{1}{3}$

26. $\dfrac{15}{4} \times \dfrac{16}{5} = \dfrac{3}{1} \times \dfrac{4}{1} = \dfrac{12}{1} = 12$

27. $\dfrac{1}{3} \times \dfrac{6}{1} = \dfrac{1}{1} \times \dfrac{2}{1} = \dfrac{2}{1} = 2$

28. $\dfrac{5}{8} \times \dfrac{10}{3} = \dfrac{5}{4} \times \dfrac{5}{3} = \dfrac{25}{12} = 2\dfrac{1}{12}$

29. $\dfrac{3}{4} \times \dfrac{1}{6} = \dfrac{1}{4} \times \dfrac{1}{2} = \dfrac{1}{8}$

30. $\dfrac{10}{3} \times \dfrac{7}{3} = \dfrac{70}{9} = 7\dfrac{7}{9}$

31. $\dfrac{5}{1} \times \dfrac{3}{7} = \dfrac{15}{7} = 2\dfrac{1}{7}$

32. $\dfrac{25}{3} \times \dfrac{4}{5} = \dfrac{5}{3} \times \dfrac{4}{1} = \dfrac{20}{3} = 6\dfrac{2}{3}$

33. $6\dfrac{1}{3} + 3\dfrac{1}{4} = 6\dfrac{4}{12} + 3\dfrac{3}{12} = 9\dfrac{7}{12}$

34. $45\dfrac{3}{8} - 12\dfrac{3}{4} = 45\dfrac{3}{8} - 12\dfrac{6}{8} = 44\dfrac{11}{8} - 12\dfrac{6}{8} = 32\dfrac{5}{8}$

35. $\dfrac{3}{4} \times 4\dfrac{1}{2} = \dfrac{3}{4} \times \dfrac{9}{2} = \dfrac{27}{8} = 3\dfrac{3}{8}$

36. $12\dfrac{1}{2} \div \dfrac{5}{8} = \dfrac{25}{2} \times \dfrac{8}{5} = \dfrac{5}{1} \times \dfrac{4}{1} = 20$

37. $50 \times 1\dfrac{3}{4} = \dfrac{50}{1} \times \dfrac{7}{4} = \dfrac{25}{1} \times \dfrac{7}{2} = \dfrac{175}{2} = 87\dfrac{1}{2}$ miles

38. $1\dfrac{2}{3} \times \dfrac{1}{4} = \dfrac{5}{3} \times \dfrac{1}{4} = \dfrac{5}{12}$ acre

NOTES

NOTES

NOTES

NOTES

NOTES